K. Kovarik · Numerical Models in Groundwater Pollution

Springer
*Berlin
Heidelberg
New York
Barcelona
Hong Kong
London
Milan
Paris
Singapore
Tokyo*

Karel Kovarik

Numerical Models in Groundwater Pollution

With 76 Figures and 46 Tables

 Springer

Dr. Karel Kovarik
University of Zilina
Faculty of Civil Engineering
Komenskeho 52
01026 Zilina
SLOVAKIA

ISBN 3-540-66792-X Springer-Verlag Berlin Heidelberg New York

Library of congress Cataloging-in-Publication Data

Kovarik, Karel.
  Numerical models in groundwater pollution / Karel Kovarik.
   p. cm.
  Includes bibliographical references and index.
  ISBN 354066792X (hc.: alk. paper)
   1. Groundwater-Pollution-Mathematical models.   2. Groundwaterflow-Mathematical models.   I. Title.

TDK426.K68   2000
628.1'68 – dc21

This work is subject to copyright. All rights reserved, whether the whole or part of the material is concerned, specifically the rights of translation, reprinting, reuse of illustrations, recitation, broadcasting, reproduction on microfilm or in any other way, and storage in data banks. Duplication of this publication or parts thereof is permitted only under the provisions of the German Copyright Law of September 9, 1965, in its current version, and permission for use must always be obtained from Springer-Verlag. Violations are liable for prosecution under the German Copyright Law.

Springer-Verlag is a company in the BertelsmannSpringer publishing group.
© Springer-Verlag Berlin Heidelberg 2000
Printed in Germany

The use of general descriptive names, registered names, trademarks, etc. in this publication does not imply, even in the absence of a specific statement, that such names are exempt from the relevant protective laws and regulations and therefore free for general use.

Cover design: Design & Production, Heidelberg
Typesetting: Best-set Typesetter Ltd., Hong Kong
SPIN 10711417   30/3136 – 5 4 3 2 1 0 – Printed on acid-free paper

Preface

This book originated from my rich experience with the development of numerical modelling software and its use in hydrogeology. I have often met with the approach to a numerical model as if it were a "black box", where the essential information about the method used (its possibilities and restrictions) was lacking. This may lead to the wrong choice of method as well as to a wrong interpretation of results. A certain conservatism is also present that rejects the most up-to-date numerical methods mainly due to their more complicated mathematics. For these reasons, I set my sights on reviewing the whole group of numerical methods from the oldest (the finite differences method) to the most modern, such as the dual reciprocity boundary element method.

After the Introduction, the next two chapters discuss the basic equations of a groundwater flow and of the transport of pollutants in a porous medium, while Chapter 4 concentrates on the fundamentals of numerical mathematics. Chapters 5 and 6 study each method of numerical modelling separately. A final comparison is to be found in Chapter 7. Several practical applications of these methods are listed in Chapter 8 and, finally, Chapter 9 explains the software included in this book.

The CD inside contains the BEFLOW system which serves to demonstrate the possibilities of the BEM and the random walk methods and the UNSDIS system as examples of the solution of vertical transport of pollutants through the unsaturated zone applying FEM. Besides the actual programs, the CD includes both their source codes (prepared by means of the Visual C++ version 6.0 programming language from Microsoft Corp.) and a gallery of results from Chapter 8. The latter is constructed as an HTML page which can be viewed either by the simple browser that we include or by a professional one (e.g. the IE or Netscape).

Finally, I wish to thank Dr. M. Drahos for allowing me to use some reports of the 1st Modelling Group in Chapter 8. Moreover, I thank my sons, Karol and Michal, for the effort put in to the translation of this book and last, but not least, my wife for her patience with me during the writing of the manuscript.

Zilina, February 2000   Karel Kovarik

Contents

1	Introduction	1
2	**Basic Equations of Groundwater Flow**	**3**
2.1	Basic Principles of Hydrodynamics	3
2.1.1	Eulerian and Lagrangian Formulations	3
2.1.2	Equilibrium of Forces in Fluid	6
2.1.3	Rate of Deformation	7
2.1.4	Navier-Stokes Equations	9
2.1.5	Potential Flow	10
2.2	Flow Through the Saturated Zone	12
2.2.1	Groundwater and Its Potential	13
2.2.2	Properties of Porous Media	14
2.2.3	The Darcy's Law	16
2.2.4	Basic Equations of Groundwater Flow	19
2.2.5	Inflows and Outflows	22
2.2.6	Two-Dimensional Case	23
2.2.7	Girinski Potential	25
2.2.8	Basic Boundary and Initial Conditions	27
2.3	Flow Through the Unsaturated Zone	28
2.3.1	Retention Curve	28
2.3.2	Governing Equations of the Unsaturated Flow	30
2.3.3	Boundary and Initial Conditions	32
	References	33
3	**Basic Equations of Transport of Pollutants in Porous Media**	**35**
3.1	Miscible Displacement	35
3.1.1	Fick's Law, Coefficient of Dispersion	36
3.1.2	Governing Equation of Transport	39
3.1.3	Sorption and Decay of Pollutants	41
3.2	Transport of Immiscible Pollutants	45
3.2.1	General Law of Darcy	45
3.2.2	Basic Equation of Multiphase Flow	46
3.2.3	General Rules of Motion of Insoluble Substances in Porous Media	47
	References	48

4	**Weighted Residuals Method**	49
4.1	Basic Terms and Definitions	49
4.1.1	Spaces	49
4.1.2	Operators	50
4.1.3	Function Spaces, Base Functions	50
4.2	Weighted Residuals Method	52
4.2.1	Moments Method	53
4.2.2	Collocation Method	54
4.2.3	Galerkin Method	54
4.2.4	Ritz Method	55
4.3	Weak Solution	55
4.4	Inverse Formulation	57
4.5	Numerical Methods Used in Problems of Groundwater Hydraulics	58
	References	60
5	**Mathematical Models of Groundwater Flow**	61
5.1	Groundwater Flow in the Saturated Zone	61
5.1.1	Analytic Solution	61
5.1.2	Finite Differences Method	64
5.1.3	Finite Element Method	69
5.1.4	The Boundary Element Method	84
5.1.5	Dual Reciprocity Method (DRM)	100
5.2	Flow in Unsaturated Zone	101
5.2.1	Analytic Solutions	102
5.2.2	Finite Differences Method	102
5.2.3	Finite Element Method	103
5.2.4	Collocation Method	106
	References	107
6	**Mathematical Models of Transport of Miscible Pollutants**	109
6.1	Basic Methods	110
6.1.1	Analytic Solution	111
6.1.2	Finite Element Method	111
6.1.3	Dual Reciprocity Method	113
6.1.4	Method of Characteristic Curves	116
6.1.5	Random Walk Method	117
6.2	Equilibrium Sorption	119
6.2.1	Analytic Solution	120
6.2.2	Finite Element Method	120
6.2.3	Particle Methods	121
6.3	Non-Equilibrium Sorption	122
6.3.1	Finite Element Method	122
6.3.2	Random Walk Method	123
	References	124

7	**Comparison of Properties of All the Methods**	125
7.1	Groundwater Flow Models	125
7.2	Models of Transport of Pollution	129
8	**Examples of the Use of Models in Practice**	131
8.1	Models to Determine Groundwater Resources	131
8.2	Models to Assess Different Influences on Groundwater	131
8.2.1	Example of Zilina Locality, Slovakia	131
8.2.2	Example of Ziar Locality, Slovakia	138
8.2.3	Example of SNP Square, Bratislava, Slovakia	145
8.3	Remediation Models	153
8.3.1	Example of Chotebor, Czech Republic	153
8.3.2	Example of Airport Bratislava, Slovakia	164
	References	174
9	**Description of Software Included in This Book**	175
9.1	BEFLOW System	175
9.1.1	Program BemInp	177
9.1.2	Program BemSolve	186
9.1.3	Program BemIsol	193
9.1.4	Program BemStream	196
9.1.5	Program BemRaw	201
9.2	System UNSDIS	207
9.2.1	Basic Relations	207
9.2.2	Commands of Program	209
9.3	Examples	212
9.3.1	System BEFLOW	212
9.3.2	Program UNSDIS	217
	References	218
Subject Index		219

1
Introduction

Modelling, as a method of analysing numerous phenomena, is used in different areas of scientific research. After its development, it was soon introduced to groundwater hydraulics and has quickly become an accepted method of exploring and predicting the behaviour of a hydrogeological environment. Practically every model is in some way related to basic governing equations which describe the phenomenon studied. In the beginning, everything was based on physical modelling which had a whole scala of different methods, from direct geometrical models of water flow (e.g. Hele-Shaw analogs) to electrohydrodynamical analogs (e.g. RC models). During recent decades, these models have been the main tools of hydraulic research.

Parallel to physical models the simpler problems were solved using a direct analytic approach (later called mathematical models). The mathematics of later successful modelling techniques was already known at that time. However, due to high computational requirements, they were not put to practical use. The widespread use of computers enabled rapid development in this area and the evolution of numerous methods of mathematical modelling. Nowadays, this type of model is more frequently used. It would be more precise to call these models numerical instead of mathematical because the governing equations are solved by numerical methods.

At the dawn of the use of numerical models, they were mostly used to study a groundwater flow in porous media. Today, our environment suffers more and more from the by-products of man's industrial activities. Groundwater is one of the things that have sustained extensive damage in the past decade. Even ordinary events in our everyday life (such as solid and liquid waste dumps, careless fertilising, accidents by transport, storing or processing of chemicals) can cause pollution of soil or surface water. The pollutants spread through a covering layer and after some time reach the groundwater and pollute it. The current situation is so bad that in some cases the problem is not to find a well with a given amount of water discharge but to find one that fulfils quality requirements. Therefore, in the 1970s the main aim of numerical models changed from the study of groundwater flow to problems of the spreading of various substances in porous media. This allows us to study groundwater pollution more effectively. For this reason numerical models have become a great tool that can be used to predict spreading of pollutants and assess different remediation projects.

Mathematical modelling can be looked at from two different points of view:

The first is that of the developer of the method (often connected with the software developer) that is going to be used in the model.

The second is the view of a hydrogeologist who is going to set up the model.

This book attempts for a balanced approach to both views, which are of a fairly different nature. In the past, people interested in numerical modelling had to cope with both views. Nowadays, there are some commercial modelling programs available so it might seem that all problems in model development have already been solved. Nevertheless, it is still true that to understand the principles of modelling correctly, it is appropriate to know the fundamentals of numerical methods.

2
Basic Equations of Groundwater Flow

From the mathematical point of view, any mathematical model is, in fact, based on solving an equation (or a system of equations) that describes a phenomenon. Such an equation is called the governing equation of the specified phenomenon. It is only natural that before we come directly to models and modelling, we concentrate on the governing equations of groundwater flow.

2.1
Basic Principles of Hydrodynamics

To make this chapter complete, a passage is inserted here concerning the basic equations of fluid mechanics that describe a motion of fluids in general (i.e. not only in a porous medium). Naturally, these equations are valid also for flow through a porous medium and they will help us later to formulate the equations governing this flow. These equations are described in detail in Kolář et al. (1983) and there is an excellent short review in Connor and Brebbia (1976).

2.1.1
Eulerian and Lagrangian Formulations

The quantities which represent the fluid's motion are pressure, velocity and temperature. These quantities are functions of time and position. Coordinates of one material particle change in time as a result of its movement. At initial time ($t = 0$) the particle has coordinates a_i and, in general, its coordinates at time t are x_i. There are two ways to express the dependence of all quantities on the position of a material particle.

The Lagrangian approach represents the quantities as functions of the initial coordinates and time, so for a velocity it means

$$v = v(a_1, a_2, a_3, t), \tag{2.1}$$

and for the position of a material particle at time t the following formula can be written

$$x_i = a_i + u_i(a_1, a_2, a_3, t), \tag{2.2}$$

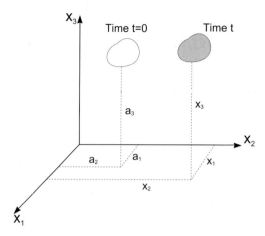

Fig. 2.1. Eulerian and Lagrangian approach

where the u_i functions are a displacement. This approach to solving problems is typical for solids where the deformations (and consequently the displacements) are small.

For fluids these deformations are generally larger. After we transform the quantities from the initial coordinates to the current ones, we have the Eulerian approach, which is based upon the fact that the current coordinates x_i and the time t are independent variables.

The Eulerian approach is more suitable for problems in hydraulics and it will be used further on (see Fig. 2.1).

Now we will concentrate on determining how fast a dependent variable changes (that means we need to determine its derivative with respect to time). Let f be a value of the studied quantity at a point (x_i, t). During a time Δt the point moves to a position $x_i + \Delta x_i$ and the quantity's value changes to $F + \Delta F$. Assuming that the function F is continuous, the following formula can be written for ΔF in a neighborhood of the point (x_i, t).

$$\Delta F = \delta F + \frac{1}{2} \delta^2 F + \ldots, \tag{2.3}$$

where

$$\delta F = \frac{\partial F}{\partial x_i} \Delta x_i + \frac{\partial F}{\partial t} \Delta t. \tag{2.4}$$

Then the limit

$$\frac{dF}{dt} = \lim_{\Delta t \to 0} \frac{\Delta F}{\Delta t} \tag{2.5}$$

2.1 Basic Principles of Hydrodynamics

is termed a substantial (or also Stokes) derivative. To express this derivative, the Taylor series is used together with an assumption that $\Delta F \cong \delta F$ with this result:

$$\lim_{\Delta t \to 0} \frac{\Delta F}{\Delta t} = \frac{\partial F}{\partial t} + \frac{\partial F}{\partial x_i} \lim_{\Delta t \to 0} \frac{\Delta x_i}{\Delta t}. \tag{2.6}$$

Components of velocity can be expressed as

$$v_i = \lim_{\Delta t \to 0} \frac{\Delta x_i}{\Delta t}, \tag{2.7}$$

and taking Eq. (2.6) into account, we obtain the following formula

$$\frac{dF}{dt} = \frac{\partial F}{\partial t} + \frac{\partial F}{\partial x_i} v_i. \tag{2.8}$$

The first term is a local derivative valid for a fixed position of the point. The remaining terms represent the influence of motion and are often called convective terms.

Let us imagine an infinitesimal prism of fluid with its edges parallel with the coordinate axes and with the prism's dimensions being dx, dy, dz, respectively (see Fig. 2.2). Its volume is $dV = dx\,dy\,dz$. The mass of the fluid inside

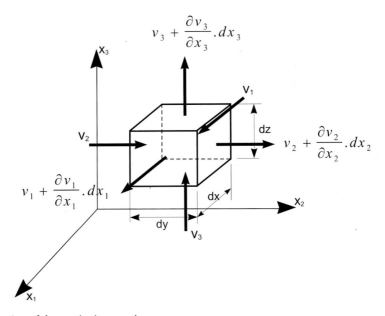

Fig. 2.2. Derivation of the continuity equation

the prism is $dM = \rho \, dV$ where ρ is the fluid's density and can generally be a function of time and position. The total mass in a volume V is determined as

$$M = \int_V \rho \, dV. \tag{2.9}$$

The net rate of a mass flow into the volume V has to be the same as the rate of production of the mass (if there is a source in the volume). It can be written as

$$\frac{dM}{dt} = \frac{d}{dt}\int_V \rho \, dV = \int_V \rho q \, dV, \tag{2.10}$$

where q is the rate of production of the mass in the volume V.

It is often supposed that there is no mass production in the given volume V. Using this assumption, Eq. (2.10) can be transformed into

$$\frac{d}{dt}\int_V \rho \, dV = 0. \tag{2.11}$$

The Reynolds transport theorem (see Connor and Brebbia 1976) implies that

$$\frac{d}{dt}\int_V \rho \, dV = \int_V \left(\frac{d\rho}{dt} + \rho \frac{\partial v_i}{\partial x_i}\right) dV = 0. \tag{2.12}$$

As the volume V is set arbitrarily, it is obvious that the next equation is true:

$$\frac{d\rho}{dt} + \rho \frac{\partial v_i}{\partial x_i} = 0. \tag{2.13}$$

If we substitute Eq. (2.8) for the substantial derivative, we have

$$\frac{\partial \rho}{\partial t} + \frac{\partial (\rho v_i)}{\partial x_i} = 0. \tag{2.14}$$

In the literature, this equation is often called the mass continuity equation. Neglecting the fluid's compressibility and assuming its density is constant, Eq. (2.14) turns into:

$$\frac{\partial v_i}{\partial x_i} = 0. \tag{2.15}$$

2.1.2
Equilibrium of Forces in Fluid

Let us define a volume V with a surface S in fluid. According to Newton's second principle of motion, the resultant of all forces acting on the volume is equal to an inertial force. There are two components of the forces that act on the fluid, volumetric forces (for example the fluid's gravity is a typical representative of

2.1 Basic Principles of Hydrodynamics

these forces) and surface forces on the surface S (such as action of the surrounding fluid on the given volume). We can write the basic equation of forces' equilibrium in the following form

$$\int_V \rho \frac{d\mathbf{v}}{dt} dV = \int_V \rho \mathbf{b} dV + \int_S \mathbf{p n} dS, \tag{2.16}$$

where there is the inertial force of the fluid's volume on the left side and on the right there is a sum of volumetric and surface forces.

Using the Gauss-Ostrograski theorem, the surface integral on the right side can be transformed into a volumetric integral. So then we obtain the equation:

$$\int_V \rho \frac{dv_i}{dt} dV = \int_V \rho b_i dV + \int_V \frac{\partial \sigma_{ij}}{\partial x_j} dV. \tag{2.17}$$

The volume is again set arbitrarily, so the previous equation changes to:

$$\rho \frac{dv_i}{dt} = \rho b_i + \frac{\partial \sigma_{ij}}{\partial x_j}. \tag{2.18}$$

On the left side partial derivatives can be substituted for the substantial derivative and we have a new equation:

$$\frac{\partial}{\partial x_j}(\rho v_i v_j) + \frac{\partial}{\partial t}(\rho v_i) = \rho b_i + \frac{\partial \sigma_{ij}}{\partial x_j}. \tag{2.19}$$

This equation is in literature often referred to as the Euler equation of equilibrium of forces in a fluid's flow.

Stress in fluid is comprised of components of pressure and viscosity

$$\sigma_{ij} = -p\delta_{ij} + \tau_{ij}. \tag{2.20}$$

If we substitute this formula into the equation of equilibrium [Eq. (2.19)], we will get a different form

$$\frac{\partial}{\partial x_j}(\rho v_i v_j) + \frac{\partial}{\partial t}(\rho v_i) = \rho b_i - \frac{\partial p}{\partial x_i} + \frac{\partial \tau_{ij}}{\partial x_j}. \tag{2.21}$$

2.1.3 Rate of Deformation

Fluids show an inner resistance to motion causing a deformation similar to solids. The definition of strain ε is the same as by solid bodies. The strain can be of two types:

- extensional strain-a relative change in length of a line segment,
- shearing strain-a change in an angle between two orthogonal line segments.

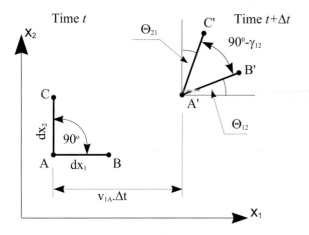

Fig. 2.3. Shear strain in fluids

In hydraulics we use the Eulerian approach and so we are going to concentrate on the change of deformation in time that means on the rate of deformation. The extensional strain rate is defined as a substantial derivative of the length dL of a line segment, divided by its length

$$\varepsilon' = \frac{1}{dL}\frac{d}{dt}(dL) \qquad (2.22)$$

or as well

$$\varepsilon' = \lim_{\Delta t \to 0}\left(\frac{\Delta(dL)}{dL.\Delta t}\right) = \lim_{\Delta t \to 0}\frac{\Delta \varepsilon}{\Delta t}. \qquad (2.23)$$

Figure 2.3 shows the initial state (at time t) and the deformed state (at time $t + \Delta t$) of two orthogonal line elements. When the velocity function is expanded in the Taylor series up to the first order, we obtain:

$$dv_i = \frac{\partial v_1}{\partial x_1}dx_1 + \frac{\partial v_1}{\partial x_2}dx_2 + \frac{\partial v_1}{\partial x_3}dx_3. \qquad (2.24)$$

The extensional strain rate of an element in the x-axis direction is

$$\varepsilon'_1 = \lim_{\Delta t \to 0}\frac{\Delta \varepsilon_1}{\Delta t} = \frac{\partial v_1}{\partial x_1}. \qquad (2.25)$$

We can obtain the strain rates in the y-axis and the z-axis directions similarly to the x-axis direction. A volumetric strain rate is a simple sum of all three values

$$\varepsilon'_V = \varepsilon'_1 + \varepsilon'_2 + \varepsilon'_3 = \frac{\partial v_1}{\partial x_1} + \frac{\partial v_2}{\partial x_2} + \frac{\partial v_3}{\partial x_3} = \frac{\partial v_i}{\partial x_i} = div\mathbf{v}. \qquad (2.26)$$

The shearing strain is represented by a change of an angle between two elements (see Fig. 2.3). This change is a sum of angular displacements θ_{12} and θ_{21}. The shearing strain rate can then be expressed similarly to the extensional strain

$$\gamma'_{12} = \lim_{\Delta t \to 0} \left(\frac{\Delta \gamma}{\Delta t} \right) = \frac{d}{dt}(\Theta_{12} + \Theta_{21}) = \frac{\partial v_1}{\partial x_2} + \frac{\partial v_2}{\partial x_1}. \tag{2.27}$$

There are two tensors that are usually defined, a symmetric one

$$e_{ij} = \frac{1}{2}\left(\frac{\partial v_i}{\partial x_j} + \frac{\partial v_j}{\partial x_i} \right), \tag{2.28}$$

and a skew-symmetric one

$$\Omega_{ij} = \frac{1}{2}\left(\frac{\partial v_j}{\partial x_i} - \frac{\partial v_i}{\partial x_j} \right). \tag{2.29}$$

Using these tensors, the deformation rate can be written in the following form

$$\varepsilon'_i = e_{ii} \quad \gamma'_{ij} = e_{ij} + e_{ji} = 2e_{ij} \quad \Theta'_{ij} = \Omega_{ji} - \Omega_{ij} \tag{2.30}$$

2.1.4
Navier-Stokes Equations

After we have derived the formulas for the rate of deformation in fluids, we can now focus on the constitutional equations. These equations describe relationships between the stress in fluid and the deformation rate. Most fluids belong to the Newtonian fluids, which are characterised by a linear relationship between the stress and the deformation

$$\tau = \mathbf{D}\mathbf{e}, \tag{2.31}$$

where

$$\begin{aligned}\mathbf{e} &= \{e_{11}, e_{22}, e_{33}, 2e_{12}, 2e_{23}, 2e_{31}\} \\ \boldsymbol{\tau} &= \{\tau_{11}, \tau_{22}, \tau_{33}, \tau_{12}, \tau_{23}, \tau_{33}\}.\end{aligned} \tag{2.32}$$

In the case of an isotropic medium, the matrix **D** has two independent elements and Eq. (2.31) can be written as

$$\tau_{ij} = \lambda e_v \delta_{ij} + 2\mu e_{ij}, \tag{2.33}$$

where μ is a dynamic viscosity coefficient and λ is a coefficient of a so-called second viscosity.

For incompressible fluids $e_v = 0$ and so we obtain a well-known formula for the viscosity of Newtonian fluids

$$\tau_{ij} = 2\mu e_{ij} = \mu\left(\frac{\partial v_i}{\partial x_j} + \frac{\partial v_j}{\partial x_i} \right). \tag{2.34}$$

We should emphasise that this formula is exact only for incompressible fluids. In the next steps we focus only on this type of fluid. When we substitute Eq. (2.34) into the Euler equilibrium equation [Eq. (2.21)], we have

$$\frac{\partial}{\partial x_j}(\rho v_i v_j) + \frac{\partial}{\partial t}(\rho v_i) = \rho b_i - \frac{\partial p}{\partial x_i} + \frac{\partial}{\partial x_j}\left[\mu\left(\frac{\partial v_i}{\partial x_j} + \frac{\partial v_j}{\partial x_i}\right)\right] \tag{2.35}$$

In the case of an isothermal flow (that means the fluid's temperature does not change) the dynamic viscosity coefficient can be considered constant and Eq. (2.35) can be transformed into

$$\frac{\partial}{\partial x_j}(\rho v_i v_j) + \frac{\partial}{\partial t}(\rho v_i) = \rho b_i - \frac{\partial p}{\partial x_i} + \mu \frac{\partial^2 v_i}{\partial x_j \partial x_j}. \tag{2.36}$$

An assumption of fluid's incompressibility implies that ρ is constant and we obtain the Navier-Stokes equation that describes the flow of the Newtonian incompressible fluids

$$\frac{\partial}{\partial x_j}(v_i v_j) + \frac{\partial v_i}{\partial t} = b_i - \frac{1}{\rho}\frac{\partial p}{\partial x_i} + v\frac{\partial^2 v_i}{\partial x_j^2}, \tag{2.37}$$

where the dynamic viscosity coefficient μ is substituted for by a kinematic viscosity coefficient $v = \frac{\mu}{\rho}$.

2.1.5
Potential Flow

The basic equations that describe the motion of water can be easily simplified if we use an assumption that the vector field of velocity is derivable from a scalar potential. That means there exists a function $\varphi(x_i)$ so that

$$v_i = \frac{\partial \varphi}{\partial x_i} \tag{2.38}$$

This function is called a scalar potential of the vector field of velocity (or a *velocity potential*).

A necessary and sufficient condition for the flow to be potential (existence of the scalar potential) is

$$\text{curl } \mathbf{v} = 0 \tag{2.39}$$

so the flow has to be irrotational. Condition (2.39) written in components is as follows

$$\frac{\partial v_i}{\partial x_j} - \frac{\partial v_j}{\partial x_i} = 0. \tag{2.40}$$

The equation of continuity for an ideal incompressible fluid [Eq. (2.15)] can be written as

2.1 Basic Principles of Hydrodynamics

$$\frac{\partial^2 \varphi}{\partial x_i^2} = 0, \qquad (2.41)$$

and we obtain the Laplace equation as the governing equation of the potential flow. If every velocity vector is always parallel with a certain plane and moreover the vectors are identical in all points of a line perpendicular to the plane, the flow is planar. In that case the velocity component orthogonal to the plane equals zero ($v_z = 0$). The flow can now be characterised by two systems of lines. The first system consists of equipotential lines (lines of the same velocity potential). The second system consists of streamlines. The streamlines are envelope curves of the velocity vector. This means that in every point of the streamline the velocity vector is tangential to the streamline. Streamlines, similarly to the equipotential lines, can be defined as lines connecting points with the same value of a function. The function is called the flow function and its symbol is ψ.

The equation of streamline is $\psi(x,y) = $ const. After a derivation thereof, we have

$$d\psi = \frac{\partial \psi}{\partial x} dx + \frac{\partial \psi}{\partial y} dy = 0. \qquad (2.42)$$

The definition of a streamline as an envelope curve of the velocity vector implies (see Fig. 2.4)

$$\frac{v_y}{v_x} = \frac{dy}{dx} \quad -v_y dx + v_x dy = 0. \qquad (2.43)$$

After comparing Eq. (2.42) and (2.43), we get

$$v_x = \frac{\partial \psi}{\partial y} \quad v_y = -\frac{\partial \psi}{\partial x}. \qquad (2.44)$$

From the definition of velocity potential it follows that

$$\frac{\partial \varphi}{\partial x} = \frac{\partial \psi}{\partial y} \quad \frac{\partial \varphi}{\partial y} = -\frac{\partial \psi}{\partial x}. \qquad (2.45)$$

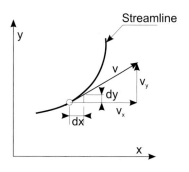

Fig. 2.4. Streamline in potential flow

We see that the potential and flow function fulfil the Cauchy-Riemann conditions and thus the functions are associated. These equations also imply

$$\frac{\partial^2 \varphi}{\partial x^2} + \frac{\partial^2 \varphi}{\partial y^2} = 0 \quad \frac{\partial^2 \psi}{\partial x^2} + \frac{\partial^2 \psi}{\partial y^2} = 0. \quad (2.46)$$

So the potential φ and the flow function ψ are also harmonic.

The equipotential lines as the lines of constant potential are described by an equation $\varphi(x,y) = \text{const}$. After its derivation, we gain

$$d\varphi = \frac{\partial \varphi}{\partial x} dx + \frac{\partial \varphi}{\partial y} dy = v_x dx + v_y dy = 0. \quad (2.47)$$

An angular coefficient of a tangent to the equipotential line (see Hálek and Švec 1979) is

$$k_\varphi = \frac{dy}{dx} = -\frac{v_x}{v_y}. \quad (2.48)$$

An angular coefficient of a tangent to the streamline in the same point is

$$k_\psi = \frac{v_y}{v_x}. \quad (2.49)$$

The angle included between these two lines can be determined from

$$\tan \alpha = -\frac{k_\varphi - k_\psi}{1 + k_\varphi k_\psi} = \infty. \quad (2.50)$$

Both tangents are orthogonal to each other.

As we mentioned above, both the functions φ and ψ are associated harmonic functions. Hence, a new complex function can be introduced. It is

$$w = \varphi(x, y) + i\psi(x, y). \quad (2.51)$$

This function is termed the complex potential of the flow. This function's region of analyticity is the whole domain of the flow except for singular points. By deriving it, we obtain a complex velocity of the flow which is defined as

$$\frac{dw}{dz} = v_x - iv_y. \quad (2.52)$$

The complex velocity is analytic as well.

2.2
Flow Through the Saturated Zone

In this section we will focus on groundwater flow through a porous medium. It is obvious that all the general equations of fluid flow are valid for groundwater flow as well. This flow, with its interaction with the porous medium, has some specific attributes. There are two basic zones to which we shall pay attention:

- *the saturated zone*, where all pores are filled with one kind of fluid (water),
- *the unsaturated zone*, where one part of the pores is filled with air.

Basic relationships in these two zones differ slightly and therefore we are going to discuss the basic governing equations of these two zones separately. Now it would be appropriate to say that the flow through the unsaturated zone is only a special case of an immiscible flow through a porous medium.

2.2.1
Groundwater and Its Potential

The term groundwater generally means all the water found beneath the Earth's surface. Groundwater can be divided into several groups according to the forces acting on it.

- *Crystallic water* is water tightly bound by physical and chemical forces in every grain of soil. This type of water cannot be removed by drying even at 105°C.
- *Pellicular water* is bound by molecular forces to the surface of soil particles. This type of water can be removed by drying at 105°C but does not move and is not available to plants.
- *Capillary water* is held by a surface tension in a form of continuous film around the soil particles. This type of water is moved by actions of a capillary force and it is available to plants.
- *Gravitational water* moves under the influence of gravity and fills all the interconnected pores in a porous medium. Most of the equations given here were derived and are valid for this type of groundwater. To underline this fact, sometimes only the gravitational water is called groundwater and the remaining types of water are then called subsurface or soil water.

The use of a macroscopic approach to modelling does not allow us to study the influence of every single force acting on groundwater. As a result, the effects of an action of these forces are represented in the form of potentials that characterise a total action of these forces. Various definitions of soil water's potential can be found in specialised works mainly concerning water flow in the unsaturated zone (see Bolt and Frissel 1960). Here, we will use only the final definition of the potential as a sum of potentials

$$\Phi = \Phi_g + \Phi_p + \Phi_o, \tag{2.53}$$

where Φ is a total potential, Φ_g is a gravity potential (from the action of gravity), Φ_p is a pressure potential (from the pressure forces), Φ_o is an osmotic potential that represents forces caused by a different chemical composition of the soil water. The pressure potential is sometimes defined as the sum of a moisture, a pneumatic and a ballast potential.

For modelling purposes, only the first two terms of Eq. (2.53) are considered and the osmotic potential is neglected.

The potential is often given in various units which are as follows:

- the basic meaning of potential is energy per a mass unit so its unit is $(J\,kg^{-1})$
- if we neglect the compressibility of water, the mass unit can be substituted for by a unit of volume and so the potential's unit is identical with a unit of pressure $(J\,m^{-3} = N\,m^{-2} = Pa)$. This form of potential is in literature often spoken of as pressure (e.g. the piezometric pressure)
- we can also express the potential as energy per a unit of gravity. We again neglect the compressibility of water and make an assumption of a constant fluid's density. Then the potential's unit is identical with a unit of length (m). This form of potential is often referred to as height (e.g. a piezometric head) and it is the most common form of potential used in practice.

2.2.2
Properties of Porous Media

Every real porous medium is a mixture of particles that originated either from weathered rocks or from remains of extinct out organisms (mainly near the surface). A porous medium is not a compact matter but a system of different-sized particles. Void spaces between these particles (pores) are randomly distributed (mostly connected with each other and filled with air, water or other fluids such as oil). The pores form a very complicated system which has to be simplified for a further study because its precise description is practically impossible. The general theory of porous media is fairly large and is beyond the scope of this book. Anyone interested in this theory may find a detailed discussion in Bear (1972).

In the theory of porous media we will apply the criteria of the theory of continuum on a porous medium. There is an interesting definition of a porous medium with the help of a representative elementary volume (REV). According to this definition, the media that include only a limited amount of unconnected void spaces are not considered porous.

The size of pores is naturally an attribute that directly influences the hydraulic properties of a medium. Now we will replace the actual medium by a fictitious continuum, which can be characterised by macroscopic quantities. The basic quantity is the volumetric porosity defined as a ratio of the volume of pores V_p to the total volume V (e.g. the volume of a sample of soil)

$$m = \frac{V_p}{V}. \qquad (2.54)$$

Although this quantity is easily determined, it says little about the hydraulic properties of a porous medium. The problem is that the volume of pores includes types of groundwater that are either not particularly movable or do not participate in the groundwater flow at all. Hence a more frequently used quantity in hydraulic practice is the effective porosity, which is defined as the ratio of the volume of moving water V_m to the total volume of the medium

$$m_e = \frac{V_m}{V}. \tag{2.55}$$

Returning to the previous chapter, we see that the major part of moving water is the gravitational water. The ratio of the volume of gravitational water to the total volume is often called aerated porosity m_a. The volume of gravitational water is given as the volume of water that flows freely out of a fully saturated soil sample.

Another important property of a porous medium is its permeability. It is defined as the ability of a medium to let pass fluids through and it is characterised by a permeability coefficient k_p. This coefficient can be expressed as

$$k_p = \frac{m_e d^2}{32}, \tag{2.56}$$

where m_e is the effective porosity and d is the average diameter of pores.

This definition stems from the idea of a porous medium as a system of parallel pipelines that are a simplified representation of pores. To determine the average velocity of a laminar water flow in a pipeline v_p, the Poiseuille formula can be used

$$v_p = \frac{\rho g i}{32 \mu} d^2, \tag{2.57}$$

where ρ is the fluid's density, i is a hydraulic gradient, d is the diameter of a pipeline (pore) and μ is the coefficient of dynamic viscosity.

A flux through one pipeline can be expressed as $Q = S_o v_p$ where S_o is a cross-sectional area of one pore and v_p is a mean velocity in pores (porous velocity). Then the total flux through a system of pores is $Q = S_p v_p$ where S_p is a total cross-sectional area of all pores. If we make the usual assumption that the areal and volumetric porosity are equal, we can use the effective porosity to determine the cross-sectional area of pores. Then we can write the following formula for the total flux

$$Q = m_e S v_p = \frac{\rho g}{\mu} \frac{m_e d^2}{32} S i = \frac{\rho g}{\mu} k_p S i. \tag{2.58}$$

The coefficient of permeability k_p characterises only a porous medium. Its value does not depend on the kind of fluid flowing through the medium. This coefficient is given in units of area (m²). Besides the permeability coefficient, a coefficient of hydraulic conductivity k is used. It is defined as

$$k = k_p \frac{\rho g}{\mu}. \tag{2.59}$$

Then formula (2.58) can be transformed into

$$Q = \frac{\rho \cdot g}{\mu} k_p S i = k S i. \tag{2.60}$$

The hydraulic conductivity coefficient is given in units of velocity ($m\,s^{-1}$). Equation (2.60) enables us to define a fictive rate v

$$v = ki. \tag{2.61}$$

This is called a rate of filtration or sometimes even a specific discharge (see Bear 1972). Equation (2.61) is referred to as the Darcy law and it is considered the basic equation of groundwater hydraulics. The hydraulic gradient i which is used in formulas (2.60) and (2.61) relates to the gradient of pressure head from pipeline hydraulics. A connection between the rate of filtration and the porous velocity is expressed by the following equation

$$v = m_e v_p. \tag{2.62}$$

The third coefficient in groundwater hydraulics that is related to the previous two is a coefficient of transmissivity T. It is defined as a multiplication of the coefficient of hydraulic conductivity and the thickness of an aquifer B.

$$T = kB. \tag{2.63}$$

The dimension of this coefficient is ($m^2\,s^{-1}$) and it is often used to solve problems of the groundwater flow in areas where it is difficult to determine the thickness of an aquifer.

The behaviour of an aquifer on which pressure forces are applied is characterised by another property which is called the compressibility of an aquifer. It is composed of the compressibility of fluid and of a solid skeleton. The volume of pores obviously changes after the aquifer's compression. This change is thought to be approximately equal to the total change of volume of a whole layer. Knowing the coefficient of compressibility of fluid β_{fl} and the coefficient of compressibility of a porous medium without any fluid β_{pm}, we can express a relative change of the aquifer's volume by the following relation

$$\frac{\Delta V}{V} = \Delta p (m \beta_{fl} + \beta_{pm}) = \Delta p \beta^*, \tag{2.64}$$

where β^* is a coefficient of aquifer's compressibility (Pa^{-1}).

2.2.3
The Darcy's Law

Equation (2.61) defines a basic law that governs the groundwater flow through the saturated zone: this is the law of Darcy, which states that the fictive rate of filtration of the groundwater flow is directly proportional to the hydraulic gradient. A coefficient of proportionality is called a coefficient of hydraulic conductivity and has the same dimension as velocity (m s^{-1}). The position of a piezometric head, according to the Bernoulli principle, is given by a sum of an elevation and the pressure head. In groundwater flow the velocity head is neglected. If we use the groundwater potential expressed as energy per gravity unit, we find that the piezometric head's position is the same as the value of the potential. The potential function is

$$\Phi = z + \frac{p}{\rho g}. \tag{2.65}$$

The current hydraulic gradient can be substituted for by the value of a negative derivative of the potential function. The Darcy's law for a one-dimensional flow has the form:

$$v = -k \frac{\partial \Phi}{\partial x}. \tag{2.66}$$

After transition to three dimensions, a piezometric surface is substituted for the piezometric line. Darcy's law then changes to

$$\mathbf{v} = -k \operatorname{grad} \Phi. \tag{2.67}$$

Equation (2.67) can be also written as

$$v_i = -k \frac{\partial \Phi}{\partial x_i}. \tag{2.68}$$

All the equations we have used until now suppose that the porous medium is isotropic. This means that the value of the hydraulic conductivity coefficient is the same in all directions. That is, however, generally not true. In most cases the medium is orthotropic – every layer of a rock medium has two coefficients of hydraulic conductivity, a maximal and a minimal one. If the orientation of these layers is matched with the coordinate axes' orientation, the Darcy's law transforms into

$$v_i = -k_{ii} \frac{\partial \Phi}{\partial x_i}. \tag{2.69}$$

If the layer is positioned so that it includes an oblique angle α with the x-axis (see Fig. 2.5), we have to transform the values of the coefficient of hydraulic conductivity (in a two-dimensional case) according to

$$\mathbf{K} = \mathbf{R}^T \mathbf{K}' \mathbf{R}, \tag{2.70}$$

where

$$\mathbf{R} = \begin{bmatrix} \cos\alpha & -\sin\alpha \\ \sin\alpha & \cos\alpha \end{bmatrix} \quad \mathbf{K}' = \begin{bmatrix} k_{max} & 0 \\ 0 & k_{min} \end{bmatrix}, \tag{2.71}$$

and we obtain

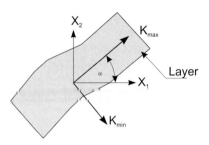

Fig. 2.5. Layer orientation

$$\mathbf{K} = \begin{bmatrix} k_{max}\cos^2\alpha + k_{min}\sin^2\alpha & (k_{min}-k_{max})\cos\alpha\sin\alpha \\ (k_{min}-k_{max})\cos\alpha\sin\alpha & k_{max}\sin^2\alpha + k_{min}\cos^2\alpha \end{bmatrix}. \tag{2.72}$$

We used a system of parallel capillaries as a model of porous medium to derive the relationships that resulted in the Darcy's law (see Section 2.2.2). Here the Darcy-Poisseuille formula for the flow in one pipe was used with the assumption of a laminar flow. Darcy's law can also be derived using the Navier-Stokes equations for a porous medium (see Connor and Brebbia 1976). To do so we have to neglect non-linear terms that have their origin in inertial forces. This implies that the validity of this law is limited.

An increasing velocity causes deviations from this law not only because of turbulence but also because of an increase of the non-linear terms. The Reynolds number is used to determine the character of the flow

$$\mathrm{Re} = \frac{vd\rho}{\mu}, \tag{2.73}$$

where μ is the dynamic viscosity of a fluid and d stands for a given element of length. Various authors differ on setting the element and on setting the critical value of the Reynolds number as well. For example in Pavlovskij (1956) it is

$$\mathrm{Re} = \frac{1}{0{,}75m + 0{,}23} \frac{vd_{10}\rho}{\mu}, \tag{2.74}$$

where d_{10} is an effective diameter of grains and Re > 1 is taken for the critical value.

For a turbulent flow the Darcy's law is substituted for by the Forchheimer formula (see Scheidegger 1960)

$$-\frac{\partial \Phi}{\partial x} = av + bv^2, \tag{2.75}$$

where a and b are constant and $a \gg b$.

The validity of the Darcy's law is limited not only by a turbulent zone but it is also limited in a prelinear regime when the velocity of the flow is too low (see Fig. 2.6). An interaction of the gravitational water with low velocity with

Fig. 2.6. Validity of Darcy's law

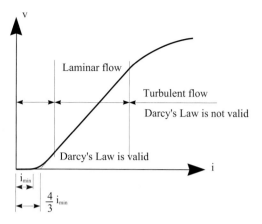

other types of groundwater (e.g. pellicular water) radically influences the flow. This non-Darcy flow is described by empirical equations (see Kutilek 1984). The simplest assumption is that there has to be a minimal hydraulic gradient i_{min} to bring water in pores to motion. The relationship then looks like this:

$$v = k\left(i - \frac{4}{3}i_{min}\right). \tag{2.76}$$

If the coefficient of hydraulic conductivity is bigger than $1 \times 10^{-7}\,\mathrm{m\,s^{-1}}$, the i_{min} can be neglected.

2.2.4
Basic Equations of Groundwater Flow

Groundwater flow is always considered to be a potential flow. That means that a potential function Φ exists. The fact that the flow is potential is equivalent to a condition that

$$\mathrm{curl}\,\mathbf{v} = 0 \tag{2.77}$$

so the flow has to be irrotational. The condition (2.77) written in components is

$$\frac{\partial v_i}{\partial x_j} - \frac{\partial v_j}{\partial x_i} = 0. \tag{2.78}$$

Here, the velocity components are actually components of the fictive rate of filtration.

A groundwater flow through the saturated zone can then be divided into two basic groups

- a flow with a phreatic (free) surface, where a groundwater level is in touch with a capillary fringe and with an aeration zone. Thus, pressure on the

surface equals the atmospheric pressure and the groundwater level is also identical with the piezometric head.
- a flow with a confined surface, where an aquifer is closed between two impervious layers and so everywhere in the aquifer pressure is higher than the atmospheric pressure.

These groups have totally different governing equations. The governing equation of the flow with a confined surface is linear and the equation of the other flow is non-linear.

The equation of continuity in porous media stems from a general equation of continuity but it contains the medium's porosity

$$\frac{d}{dt}\int_V \rho m\, dV = \int_V \left(\frac{d(\rho m)}{dt} + \frac{\partial(\rho m v_{pi})}{\partial x_i} \right) dV = 0, \qquad (2.79)$$

where m is the porosity and v_p is the actual (porous) velocity. Equation (2.79) is again valid for an arbitrary volume and so

$$\frac{\partial(\rho m)}{\partial t} + \frac{\partial(\rho m v_{pi})}{\partial x_i} = 0. \qquad (2.80)$$

If we substitute the rate of filtration for the multiplication of the porous velocity and the porosity, we obtain

$$\frac{\partial}{\partial t}(\rho m) + \frac{\partial}{\partial x_i}(\rho v_i) = 0. \qquad (2.81)$$

This is the equation of continuity for porous media. If the rate of filtration is replaced using the Darcy's law, Eq. (2.81) changes to

$$\frac{\partial}{\partial t}(\rho m) - \frac{\partial}{\partial x_i}\left(\rho k_{ii} \frac{\partial \Phi}{\partial x_i} \right) = 0. \qquad (2.82)$$

Let us assume that the density ρ and the porosity m are constant. This, along with an assumption of the fluid's and the medium's incompressibility, alters the previous equation to

$$\frac{\partial}{\partial x_i}\left(k_{ii} \frac{\partial \Phi}{\partial x_i} \right) = 0, \qquad (2.83)$$

which is a basic governing equation of a steady groundwater flow. Presuming that k_{ii} is constant, the equation above has the form

$$k_{ii} \frac{\partial^2 \Phi}{\partial x_i^2} = 0. \qquad (2.84)$$

An appropriate transformation of coordinates transforms the previous equation into the Laplace equation. A usual coordinates' transformation is

$$\tilde{x}_i = x_i \sqrt{\frac{k_0}{k_{ii}}}, \qquad (2.85)$$

where k_0 is generally an arbitrary constant that has the same dimension as velocity (m s^{-1}). This constant is often set to have the same value as the coefficient of hydraulic conductivity in the z-axis direction ($k_0 = k_{33}$). After applying it in Eq. (2.84), we have

$$\frac{\partial^2 \Phi}{\partial (\tilde{x}_i)^2} = 0. \tag{2.86}$$

In some cases it is not suitable to neglect the fluid's and medium's compressibility, and so we have to use the equation of continuity in the form of Eq. (2.14). Then the left side of this equation can become

$$\frac{\partial}{\partial t}(\rho m) = \rho \frac{\partial m}{\partial t} + m \frac{\partial \rho}{\partial t}, \tag{2.87}$$

and after the application of the Hooke law, we obtain

$$\frac{\partial \rho}{\partial t} = \rho \beta_{kv} \frac{\partial p}{\partial t} \quad \frac{\partial m}{\partial t} = \beta_{pm} \frac{\partial p}{\partial t}. \tag{2.88}$$

β_{fl} is the coefficient of fluid compressibility and β_{pm} is the coefficient of medium compressibility. Using the potential's definition, a derivative of pressure with respect to time can be expressed as

$$\frac{\partial p}{\partial t} = \rho g \frac{\partial \Phi}{\partial t} = \gamma \frac{\partial \Phi}{\partial t}. \tag{2.89}$$

A substitution in Eq. (2.87) changes it to

$$\gamma \rho \beta^* \frac{\partial \Phi}{\partial t} - \frac{\partial}{\partial x_i}(\rho v_i) = 0. \tag{2.90}$$

The second term of this equation is usually arranged to the following form

$$\frac{\partial}{\partial x_i}(\rho v_i) = \rho \frac{\partial v_i}{\partial x_i} + v_i \frac{\partial \rho}{\partial x_i}. \tag{2.91}$$

Considering the fact that the velocity of a flow in a porous medium alongside the density's derivative is very small, any term that contains both of them can be neglected. Then Eq. (2.91) is simplified to

$$\gamma \beta^* \frac{\partial \Phi}{\partial t} - \frac{\partial v_i}{\partial x_i} = 0. \tag{2.92}$$

If we substitute Eq. (2.69) for the rate of filtration, we obtain

$$\gamma \beta^* \frac{\partial \Phi}{\partial t} - \frac{\partial}{\partial x_i}\left(k_{ii} \frac{\partial \Phi}{\partial x_i}\right) = 0. \tag{2.93}$$

This is the equation governing groundwater flow with a confined surface. The term $\gamma\beta^*$ is often called a specific coefficient of storativity S_s with a dimension (m^{-1}).

2.2.5
Inflows and Outflows

When deriving Eq. (2.79) and (2.93), we paid no attention to the outflow of groundwater from an aquifer (or the inflow into an aquifer). A transformation of the basic equation describing the rate of change of volume to the relation below is a usual practice in hydrogeological modelling

$$\int_V \left(\frac{d(\rho m)}{dt} + \frac{\partial(\rho m v_{pi})}{\partial x_i} \right) dV = \int_V \rho m q \, dV, \tag{2.94}$$

and the governing equation of the groundwater flow has a new form

$$\gamma\beta^* \frac{\partial \Phi}{\partial t} - \frac{\partial}{\partial x_i}\left(k_{ii} \frac{\partial \Phi}{\partial x_i} \right) - q = 0. \tag{2.95}$$

Here, q is a specific discharge per a unit of volume of an aquifer and it has a dimension (s^{-1}).

A hydrogeological borehole in an aquifer with a yield $Q(t)$ is represented by a point source with a specific yield

$$q(t) = \frac{Q(t)}{bF}, \tag{2.96}$$

where F is an area of the borehole's cross-section which is orthogonal to its axis and b is a length of the borehole's filter.

Let us presume that the borehole's axis is vertical and one of its points has the coordinates x_{0i}. Moreover, we assume that the borehole's radius converges to zero. So we consider the borehole to be a linear point source that fulfils these conditions (see Bansky and Kovarik 1978): $q = 0$ when $x_i \neq x_{0i}$, $q \to \infty$ when $x_i = x_{0i}$ and

$$\int_{-\infty}^{\infty}\int_{-\infty}^{\infty} bq\,dx_i = Q. \tag{2.97}$$

It is obvious that these conditions can be expressed using the Dirac function in the form below

$$q = \frac{Q}{b}\delta(x_i - x_{0i}). \tag{2.98}$$

After substitution into Eq. (2.95), we gain

$$\gamma\beta^* \frac{\partial \Phi}{\partial t} - \frac{\partial}{\partial x_i}\left(k_{ii} \frac{\partial \Phi}{\partial x_i} \right) - \frac{Q}{b}\delta(x_i - x_{0i}) = 0. \tag{2.99}$$

If we have more point sources, we use a superposition principle and we add up the effects of the sources.

2.2.6
Two-Dimensional Case

Hitherto, we have always solved problems in a three-dimensional space. It is obvious that some problems can be solved more easily in two dimensions. Later in the case studies we will see that also a two-dimensional model can be set up and solved far more easily. The transformation of a three-dimensional groundwater flow to two dimensions was done for the first time by Dupuit, and is thus called the Dupuit theorem.

Let us define a coordinate system x,y,z so that the z-axis is perpendicular to an equipotential plane where the potential equals zero and the origin of coordinates is in the same plane. The basic equation of a three-dimensional flow of an incompressible fluid can be written as

$$\frac{\partial}{\partial x}\left(k_x \frac{\partial \Phi}{\partial x}\right) + \frac{\partial}{\partial y}\left(k_y \frac{\partial \Phi}{\partial y}\right) + \frac{\partial}{\partial z}\left(k_z \frac{\partial \Phi}{\partial z}\right) + q = 0. \tag{2.100}$$

$\Phi = z$ is true for any point of the phreatic surface.

Let us assume that the value of the coefficient k_z is infinite. Equation (2.100) is valid in the whole area of filtration (not only in singular points where the rate v_z can be infinite) only if the potential's derivative in the z-axis direction approaches zero, i.e. $\dfrac{\partial \Phi}{\partial z} \to 0$.

This implies that $\Phi(z) = \text{const}$. An assumption that the potential (the piezometric head) is constant in a vertical direction is called the Dupuit assumption. It allows us to transform the governing equation of a three-dimensional groundwater flow to two dimensions. When Eq. (2.100) is integrated over the z-coordinate on an interval that is given by the thickness of an aquifer, it acquires a new form

$$\int_{z_0}^{z_1}\left[\frac{\partial}{\partial x}\left(k_x \frac{\partial \Phi}{\partial x}\right) + \frac{\partial}{\partial y}\left(k_y \frac{\partial \Phi}{\partial y}\right) + q\right]dz + \int_{z_0}^{z_1}\frac{\partial}{\partial z}\left(k_z \frac{\partial \Phi}{\partial z}\right)dz = 0. \tag{2.101}$$

The point z_0 lies on the bottom boundary and the point z_1 lies on the top boundary of the aquifer.

If it is a flow with a phreatic surface, the point z_0 is on the surface of an impermeable subsoil and the point z_1 is on the phreatic surface. According to the Dupuit assumption, the second term in Eq. (2.101) equals zero and so its integral is a constant. This constant is very often expressed as a subtraction of two rates (a rate of infiltration into the phreatic surface v_0 and a rate of change of the surface's position v_1). Then we have

$$\frac{\partial}{\partial x}\left[k_x(z_1-z_0)\frac{\partial \Phi}{\partial x}\right]+\frac{\partial}{\partial y}\left[k_y(z_1-z_0)\frac{\partial \Phi}{\partial y}\right]+q(z_1-z_0)-(v_1-v_0)=0. \qquad (2.102)$$

$B = z_1 - z_0$ is the thickness of an aquifer. Pressure on the phreatic surface is the same as the atmospheric pressure and so $\Phi = h$.

The following formula is valid for the rate of change of the phreatic surface

$$v_1 = m_e \frac{\partial h}{\partial t}. \qquad (2.103)$$

After substitution of Eq. (2.103) into Eq. (2.102), we obtain the Boussinesq equation

$$\frac{\partial}{\partial x}\left(k_x B \frac{\partial h}{\partial x}\right)+\frac{\partial}{\partial y}\left(k_y B \frac{\partial h}{\partial y}\right)+qB+v_0 = m_e \frac{\partial h}{\partial t}. \qquad (2.104)$$

This is the governing equation of a two-dimensional flow with a phreatic surface (an unconfined flow).

In the case of a confined flow, the point z_1 is between the aquifer and the top impermeable layer. Because the confined flow shows an elastic behaviour, we have to use Eq. (2.95). The integration is the same but the rate v_0 is not the rate of infiltration. It can, for example, be the rate of leakage in the case of a semi-permeable upper layer. The rate v_1 equals zero. Instead of Eq. (2.104) we have

$$\frac{\partial}{\partial x}\left(k_x B \frac{\partial \Phi}{\partial x}\right)+\frac{\partial}{\partial y}\left(k_y B \frac{\partial \Phi}{\partial y}\right)+qB+v_0 = S_s B \frac{\partial \Phi}{\partial t}. \qquad (2.105)$$

In contrast to the flow with a phreatic surface, the thickness B is not a function of the surface's position in the confined flow. As a result, the term containing the coefficient of hydraulic conductivity and the thickness of an aquifer is often substituted for by the coefficient of transmissivity T and the term consisting of the coefficient of specific storativity S_s and the thickness of an aquifer is substituted for by the coefficient of storativity S.

The governing differential equation of the confined flow is written as follows

$$T_x \frac{\partial^2 \Phi}{\partial x^2}+T_y \frac{\partial^2 \Phi}{\partial y^2}+qB+v_0 = S\frac{\partial \Phi}{\partial t}. \qquad (2.106)$$

If we needed to include effects of point sources (or sinks) in the previous equation, we would use the same approach to include the sources in the equation as we did in case of the 3-D problem [see Eq. (2.99)] when we used the Dirac function. If we suppose that the wells are fully penetrated (this means the filter's length b equals the thickness of the aquifer B), we obtain this equation

$$T_x \frac{\partial^2 \Phi}{\partial x^2}+T_y \frac{\partial^2 \Phi}{\partial y^2}+qB+v_0+\sum_{k=1}^{N} Q_k \delta(x-x_{0k},y-y_{0k}) = S\frac{\partial \Phi}{\partial t}. \qquad (2.107)$$

There are a number of problems where we neglect any change of the flow's potential in time. This flow is called stationary and the right side of Eq. (2.107) equals zero

$$T_x \frac{\partial^2 \Phi}{\partial x^2} + T_y \frac{\partial^2 \Phi}{\partial y^2} + qB + v_0 + \sum_{k=1}^{N} Q_k \delta(x - x_{0k}, y - y_{0k}) = 0. \quad (2.108)$$

2.2.7
Girinski Potential

The Boussinesq equation is non-linear because the thickness of an aquifer is a function of groundwater level $B = f(h)$. Therefore to solve the equation, we have to make some simplifying assumptions. One of the most frequently presumptions is that the impermeable subsoil is horizontal and the coefficients of hydraulic conductivity are constant. Then the aquifer's thickness is simply $B = h$ and Eq. (2.104) can be written as

$$\frac{\partial}{\partial x}\left(k_x h \frac{\partial h}{\partial x}\right) + \frac{\partial}{\partial y}\left(k_y h \frac{\partial h}{\partial y}\right) + Q\delta(x - x_0, y - y_0) + v_0 = m_e \frac{\partial h}{\partial t}. \quad (2.109)$$

The right side of Eq. (2.109) equals zero for the stationary flow and after some derivations it looks like this:

$$k_x h \frac{\partial^2 h^2}{\partial x^2} + k_y \frac{\partial^2 h^2}{\partial y^2} + 2Q\delta(x - x_0, y - y_0) + 2v_0 = 0. \quad (2.110)$$

where we used the relations below

$$\frac{\partial}{\partial x}\left(h \frac{\partial h}{\partial x}\right) = \frac{1}{2}\frac{\partial^2(h^2)}{\partial x^2} \quad \frac{\partial}{\partial x}\left(h \frac{\partial h}{\partial y}\right) = \frac{1}{2}\frac{\partial^2(h^2)}{\partial y^2}. \quad (2.111)$$

The previous equation now changes to the Poisson equation with h^2 as an unknown quantity. Let us suppose that $k_x = k_y = k$ and define a specific discharge using these equations

$$q_x = -kh \frac{\partial h}{\partial x} \quad q_y = -kh \frac{\partial h}{\partial y}. \quad (2.112)$$

Then Eq. (2.110) acquires this form:

$$\frac{\partial q_x}{\partial x} + \frac{\partial q_y}{\partial x} - Q\delta(x - x_0, y - y_0) - v_0 = 0. \quad (2.113)$$

We introduce the Girinski potential (see Hálek and Švec 1979; Bear 1972) as a function satisfying the two conditions listed below

$$q_x = \frac{\partial G}{\partial x} \quad q_y = \frac{\partial G}{\partial y}. \quad (2.114)$$

Equation (2.113) is transformed to

$$\frac{\partial^2 G}{\partial x^2} + \frac{\partial^2 G}{\partial y^2} - Q\delta(x-x_0, y-y_0) - v_0 = 0. \tag{2.115}$$

The Girinski potential in the case of a flow with a pheatic surface in an aquifer with a horizontal impermeable subsoil, is defined as

$$G(x,y) = \int_0^h k(z-h)dz, \tag{2.116}$$

where h is a groundwater level above the subsoil and z is the vertical coordinate. In a homogeneous isotropic medium the potential is

$$G = -k\frac{h^2}{2}. \tag{2.117}$$

The main advantage of the Girinski potential lies in its use in a multilayer medium. If there are $(n + 1)$ horizontal layers and each layer has its own hydraulic conductivity coefficient k_i and its initial and final coordinate z_{i-1} and z_i (see Fig. 2.7), the Girinski potential is

$$G(x,y) = \int_{z_0}^{z_1} k_1(z-h)dz + \sum_{i=2}^{n+1} \int_{z_{i-1}}^{z_i} k_i(z-h)dz. \tag{2.118}$$

After integration, we have (see Hálek and Švec 1979)

$$G(x,y) = -h\left[k_{n+1}h + \sum_{i=1}^{n} z_i(k_i - k_{i+1})\right] + \frac{1}{2}\sum_{i=1}^{n} z_i^2(k_i - k_{i+1}). \tag{2.119}$$

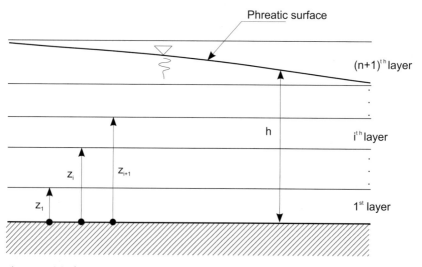

Fig. 2.7. Girinski potential

The potential in this form is often used for analytic solutions of problems in a planar groundwater flow (see Chap. 5.1).

2.2.8
Basic Boundary and Initial Conditions

The main aim of mathematical models is to solve the governing differential equations. As is known from the mathematical analysis, to solve this kind of equation it is necessary to set boundary (and sometimes also initial) conditions. A boundary condition is a general formula that is used to determine values yielded by the result (or derived values) in a zone of interest (mainly on the boundary). An initial condition is used when problems concerning an unsteady flow are solved and when it is necessary to know the result's value at the initial time.

From a mathematical point of view, the boundary conditions are divided into three basic groups

- boundary conditions of the 1st kind (the Dirichlet conditions), where the potential's value on the boundary is given.
- boundary conditions of the 2nd kind (the Neumann conditions), where the value of the potential's normal derivative on the boundary is given.
- boundary conditions of the 3rd kind (the Cauchy conditions), where the given derivative is a function of the potential.

From a hydraulic point of view, the boundary can be divided into different parts

- *an impermeable part.* There is no groundwater inflow or outflow. This part is characterised by boundary conditions of the 2nd kind with the value of the potential's normal derivative being zero. This part of the boundary is a streamline and equipotentials are perpendicular to it.
- *a permeable part.* Here the groundwater flows in and out and that makes the choice of a boundary condition difficult. This choice depends on the experience of the hydrogeologist. In this part all three types of boundary conditions can be used. If the boundary is a river or a lake bank, a condition of the 1st kind is used, that means $\Phi = f(x,y,z,t)$ where f is a known function. This condition is very strong because it presumes an ideal interaction between the groundwater surface and the water level in the river. A more precise match of the flow requires a boundary condition of the 3rd kind

$$\sum_i k_{ii} \frac{\partial \Phi}{\partial x_i} = f(x_i, t, \Phi), \qquad (2.120)$$

where f is again a known function not only of coordinates and time but of the potential as well. This condition can also be used to fit clogged river banks (for more details see Mucha and Shestakov 1987). If we can determine the quantity of water flowing into the zone through this part, it can be characterised by a boundary condition of the 2nd kind,

$$\sum_i k_{ii} \frac{\partial \Phi}{\partial x_i} = f(x_i, t). \tag{2.121}$$

- *a part with phreatic surface*. The pressure on a free surface equals the atmospheric pressure (which can be neglected)

$$\Phi = h + \frac{p}{\rho g} = h. \tag{2.122}$$

If there is neither infiltration nor evaporation, the rate perpendicular to the surface equals zero and the surface is a streamline.

2.3
Flow Through the Unsaturated Zone

Here we claim that the saturated and the unsaturated zone are separated by the free surface of the groundwater, though in reality there is no clear interface. This is caused mostly by the capillary fringe. Phenomena that occur in both zones are investigated by two separate groups of researchers, so we could say that these two zones are two separate worlds. Hydrogeology concentrates only on the saturated zone. The unsaturated zone is examined mostly by soil scientists, but their examination never reaches beyond the groundwater level. Modern numerical methods enable a complex solution of the whole problem. This chapter aims at a flow in the unsaturated zone. Anyone interested in this matter can find it in specialized books (e.g. Kutilek 1984).

In the unsaturated zone there is a three-phase system and the soil skeleton contains two phases (fluid and vapour). In terms of terminology, it is more correct to refer to transfer of moisture than to groundwater flow.

2.3.1
Retention Curve

As with other transfer phenomena, there is a condition of a force that enables the transfer of moisture as well. This force is generally a gradient of the groundwater potential, which was defined in Section 2.2. The potential used in a solution of the moisture transfer is defined as a sum of a moisture and a gravity potential

$$\Phi = \Phi_w + \Phi_g. \tag{2.123}$$

The most frequently used form of potential is expressed as energy per a unit of gravity, that is in units of length. Hence the potential is

$$\Phi = H + x_3, \tag{2.124}$$

where H is the so-called pressure head and x_3 stands for a coordinate in the direction of Earth's gravity.

The relationship between the soil's moisture and the moisture potential (or the pressure head) is called a retention curve (see Fig. 2.8). This name comes from the way in which porous media can retain water (see, for example, Kutilek 1984).

As the range of the moisture potential is fairly large, its logarithmic value is used instead. Soil scientists use a symbol pF for this logarithmic value and it is defined as

$$pF = \log \Phi_w. \tag{2.125}$$

For this reason the retention curve is sometimes called in literature a pF curve. Many different methods are used to set this curve in laboratory (for more details see Kutilek 1984).

The behaviour of a porous medium in the case of the moisture increasing (imbibition state) differs from the behaviour when the moisture decreases (drainage state). This is called a hysteresis of the retention curve, which means that the curve depends on how the equilibrium was reached (see Fig. 2.8). If we dry a sample of soil that was fully saturated with water, we obtain a primary drying branch of the retention curve (see Fig. 2.8). After reaching the minimal moisture, the sample was allowed to absorb water, and this time a dependency of the moisture on the potential is expressed by the primary wetting branch of

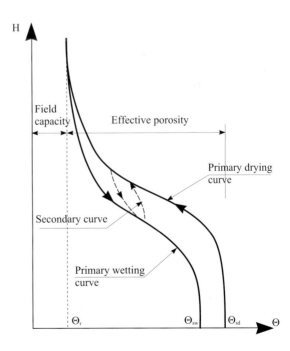

Fig. 2.8. Retention curve

the curve. Between these two branches secondary branches appear when the sample's draining-off was stopped before reaching the minimal moisture. This is the reason why a study of behaviour of porous media is so difficult when drainage and imbibition states change quickly.

A few characteristic points on the retention curve can be seen in Fig. 2.8. The primary drying branch begins at a point of a total saturation $\Theta - \Theta_s$ when the moisture potential equals zero. The branch's beginning is parallel with the H-axis and therefore $d\Theta/dH = 0$. This parallel part of the curve ends at the point A with a value H_a and this pressure head is called a *critical capillary head*. When expressed in terms of pressure, the point A is the *bubbling pressure*. From this point onwards, the moisture gradually decreases to its minimum, which is called *field capacity* (or residual moisture Θ_r). A complement to the field capacity is called the effective porosity m_e (see Section 2.2).

The mathematical model requires setting an analytic relationship $\Theta = f(H)$. There are a number of relationships and currently the most common is (van Genuchten 1981)

$$\Theta(H) = \Theta_r + (\Theta_s - \Theta_r)\left[1 + \left(\alpha|H|^n\right)\right]^{-m}, \tag{2.126}$$

where α, n are characteristic parameters of soil, $m = 1 - \dfrac{1}{n}$.

2.3.2
Governing Equations of the Unsaturated Flow

In the unsaturated zone velocity of the flow is governed by similar laws as those derived in case of the saturated zone. The formula (2.61), which says that the fictive rate of filtration is directly proportional to the hydraulic gradient, is still valid. The only difference compared to the law of Darcy is that the coefficient of proportion is not constant, but depends on the degree of the medium's saturation with water. It is a function of the soil's moisture Θ. Thus, Buckingham (1907) was the first to describe this relationship in the unsaturated zone and the equation carries the name the Darcy-Buckingham equation. In more dimensions the equation looks like this:

$$v = K(\Theta)\,\text{grad}\,\Phi. \tag{2.127}$$

The coefficient K in Eq. (2.127) is called the unsaturated hydraulic conductivity (or sometimes a capillary conductivity). It is a function of the soil's moisture and it is obvious that the coefficient is always smaller than the hydraulic conductivity coefficient which is set for a totally saturated medium. The form of this coefficient more often used is the form of a function of the potential (or the pressure head). To help us we use the dependency of moisture on the pressure head as was described in the previous chapter. The coefficient of unsaturated hydraulic conductivity clearly depends on the soil's moisture but because of the hysteresis of the retention curve there is a hysteresis in the dependency of K on the pressure head.

There are various formulas from different authors which are used in calculations. One very often used is the formula derived by van Genuchten

$$K(H) = k_s \frac{\left\{1-(\alpha|H|)^{n-1}\left[1+(\alpha|H|)^n\right]^{-m}\right\}^2}{\left[1+(\alpha|H|)^n\right]^{\frac{m}{2}}}, \qquad (2.128)$$

where k_s is the coefficient of hydraulic conductivity of a saturated medium and α, n are characteristic soil parameters [see Eq. (2.126)].

The process of deriving the basic governing equation is similar to the one we used in Section 2.2 for the saturated zone. We begin with the continuity equation (2.14)

$$\frac{\partial}{\partial t}(\rho m) + \frac{\partial}{\partial x_i}(\rho v_i) = 0, \qquad (2.129)$$

where fluid can still be considered incompressible. Ergo, ρ = constant. Because the moisture changes result in a change in the degree of pore saturation, the moisture Θ is substituted for the porosity m. Thereafter Eq. (2.129) becomes

$$\frac{\partial \Theta}{\partial t} + \frac{\partial v_i}{\partial x_i} + z = 0. \qquad (2.130)$$

A new term Z in the equation symbolises the change of water's volume in soil (e.g. water consumption of plants). After using the Darcy-Buckingham law, the equation is transformed to

$$\frac{\partial \Theta}{\partial t} - \frac{\partial}{\partial x_i}\left(K(H)\frac{\partial \Phi}{\partial x_i}\right) + Z = 0. \qquad (2.131)$$

There are three variables in this new equation-the moisture Θ, the pressure head H and the potential Φ. It is evident that this form is not suitable and some changes are needed. This equation can be derived in two ways. Consequently, there are two forms of the governing equation. The equation is often derived so that the pressure head H is the only variable. This is a *capacity form* of the governing equation

$$C_w \frac{\partial H}{\partial t} - \frac{\partial}{\partial x_i}\left(K(H)\frac{\partial H}{\partial x_i}\right) + \frac{\partial K}{\partial x_3} + Z = 0 \quad C_w = \frac{d\Theta}{dH}, \qquad (2.132)$$

where $C_w(H)$ is called the water capacity.

The second alternative is a *diffusion form* of the equation where the following formula is substituted for the potential's derivative

$$\frac{\partial \Phi}{\partial x_i} = \frac{dH}{d\Theta}\frac{\partial \Theta}{\partial x_i} + \frac{\partial x_3}{\partial x_i}. \qquad (2.133)$$

Then the moisture Θ is the only variable and the governing equation can be expressed as (see Kutilek 1984):

$$\frac{\partial \Theta}{\partial t} - \frac{\partial}{\partial x_i}\left(D(\Theta)\frac{\partial \Theta}{\partial x_i}\right) + \frac{dK}{d\Theta}\frac{\partial \Theta}{\partial x_3} + Z = 0 \quad D(\Theta) = K(\Theta)\frac{dH}{d\Theta}. \tag{2.134}$$

$D(\Theta)$ is called a capillary diffusivity because it has the same dimension as the coefficient of diffusion ($m^2 s^{-1}$) (see Chap. 3). That is the only similarity with the diffusion because the force that causes the diffusion and the one that causes the water transport are of a different nature.

2.3.3
Boundary and Initial Conditions

When solving problems of water transport in the unsaturated zone, it is again necessary to define the initial and boundary conditions of a solution. Beside the ordinary boundary conditions that resemble those in the saturated zone, there are some special conditions (see Vogel 1989). The vast majority of problems that are resolved in the unsaturated zone concern vertical water transport, being either a one-dimensional or a two-dimensional problem of water transport in a vertical profile. In numerical modelling we use the equation in the capacity form so that the only variable is the pressure head H. This head is negative in the unsaturated zone, it equals zero on the groundwater surface and it is positive below the groundwater level.

Before we start a simulation we have to know the initial value of the pressure head. This is often a problem because any measurement of pressure in pores is done only exceptionally. An easier way is to measure the soil's moisture and then to transform these values into those of the pressure head using the equation of the retention curve. An initial condition can be written in this form

$$H(x_i, t_0) = f_0(x_i). \tag{2.135}$$

Basic boundary conditions are mostly of the 1st kind (Dirichlet conditions) and for the pressure head they are as follows

$$H(x_i, t) = f_1(x_i, t) \quad x_i \in \Gamma_1. \tag{2.136}$$

Boundary conditions of the 2nd kind (Neumann conditions) are also common

$$-K\left(\frac{\partial H}{\partial x_i} + \frac{\partial x_i}{\partial x_3}\right) = f_2(x_i, t) \quad x_i \in \Gamma_2. \tag{2.137}$$

Moreover, there are some special boundary conditions (see Vogel 1989; Feddes et al. 1978). Above all, it is the condition of a pond, i.e. gathering of water on the surface. This condition is

$$-K\left(\frac{\partial H}{\partial x_i} + \frac{\partial x_i}{\partial x_3}\right) = f_3(x_i, t) - \frac{\partial H}{\partial t} \quad x_i \in \Gamma_3. \tag{2.138}$$

and it is valid only for H greater then or equal to zero. If H is negative, it means that the pond on the surface has already infiltrated and the condition changes to an ordinary inflow condition [see Eq. (2.137)].

References

Bansky V, Kovarik K (1978) Riesenie trojrozmerneho nestacionarneho prudenia podzemnej vody metodou konecnych prvkov. Vodohosp. cas. 26:293–314

Bear J (1972) Dynamics of fluids in porous media. American Elsevier, New York

Bolt GH, Frissel MJ (1960) Thermodynamics of soil moisture, Neth. J. Agric. Sci. 857–78

Buckingham E (1907) Studies on the movement of soil moisture, U.S. Dept. Agr. Bur. Soils Bull., 38, Washington DC

Connor JJ, Brebbia CA (1976) Finite element techniques in fluid flow. Butterworth & Comp., London

Feddes RA, Kowalik P, Zaradny H (1978) Simulation of field water use and crop yield. J. Wiley, New York

Hálek V, Svec J (1979) Groundwater hydraulics. Academia, Prague

Kolár V, Patocka C, Bém J (1983) Hydraulika. SNTL, Prague

Kutilek M (1984) Vlhkost pórovitých materiálu. SNTL, Prague

Mucha I, Shestakov VM (1987) Hydraulika podzemnych vod, Alfa, Bratislava

Pavlovskij NN (1956) Sobranie socinenij II, Moskva, Izd. AN SSSR

Scheidegger AE (1960) The physics of flow through porous media. 2nd edition, University of Toronto Press, Toronto

van Genuchten MT (1981) Non-equilibrium transport parameters from miscible displacement experiments. Research Report No. 119, US Salinity Lab., USDA

Vogel T (1989) SWM I-Numericky model jednorozmerného proudeni v nenasycene pudní zone. In: Nemec V, et al. (eds) Hornická Pribram ve vede a technice, CSVTS Pribram pp. 543–556

3
Basic Equations of Transport of Pollutants in Porous Media

As groundwater pollution spreads, the transport of pollutants becomes a more important part of models. In contrast to a solution of groundwater flow, the transport of pollutants is a fairly young discipline that began development in the 1960s. This is connected with the complexity of the problem and the difficult tools needed for the solution. Apart from that, the transport of pollutants is a typical interdisciplinary problem that requires the cooperation of a number of specialists, which is not always perfect. Hydrodynamics aims at geometrical aspects of dispersion, and geochemistry focuses on chemical reactions without considering the geometry. However, we still do not fully understand the phenomena that are connected with the transport of pollutants in porous media.

The transport of pollutants in porous media can be divided into two main groups

- a miscible displacement-pollutants are soluble in water
- an immiscible displacement-pollutants are insoluble in water

3.1
Miscible Displacement

The quantity of a substance in groundwater is represented by its concentration. The concentration is thought of as the substance's quantity per unit of water volume ($mg\,l^{-1}$). The transport is an irreversible, non-stationary process that creates a transition zone where the concentration continuously changes from the minimal to the maximal value. This phenomenon is called a dispersion and it is a macroscopic reflection of a real movement of particles in pores and a reflection of different physical and chemical phenomena. Generally, there are many reasons for these movements, of which the most important ones are:

- outer forces acting on fluid,
- geometry of pores,
- molecular diffusion,
- changes in the fluid's properties (i.e. density, viscosity),
- changes in the concentration caused by physical and chemical processes in fluid (such as a radioactive decay),
- interactions between the solid and the liquid phase (an adsorptive or desorptive process),

The spread of pollutants in water is caused by two mechanisms:

- The molecular diffusion that has its origin in the Brownian motion of molecules. It is the same in every direction and is independent of the velocity of a flow.
- A hydrodynamic dispersion is created by the fluid's motion in a porous medium.

There is a mechanical spread of the pollutant due to the medium's irregularity and randomly grouped pores. Additionally, water flows faster in the middle of each pore, which results in a further spread of the pollutant. This is called a microscopic dispersion, and is comparable with the molecular diffusion.

Both these mechanisms act at the same time and the one complements the other. It is obvious that molecular diffusion prevails in cases of a flow in low permeable media with low velocities. This mechanism causes the well-known effect of a dissolved pollutant's spread even in a direction opposite to the that of groundwater flow. Hydrodynamic dispersion prevails in the case of a highly permeable medium with a high velocity of flow.

3.1.1
Fick's Law, Coefficient of Dispersion

The velocity of transport can be expressed as a mass flux; that is, the mass of the pollutant which is transported through a given area per unit of time. A specific mass flux is defined, for our further use, as the quantity of the pollutant transported through a unit of area per unit of time. Molecular diffusion depends on the gradient of concentration; this means that the pollutant moves from a point with higher concentration to one with lower concentration. There is a linear proportion between the specific mass flow caused by the molecular diffusion and the concentration gradient. This can be seen in Fick's law of diffusion

$$q_M = -\Theta D_M \operatorname{grad} C, \tag{3.1}$$

where D_M is a coefficient of molecular diffusion ($m^2 s^{-1}$). In the case of transport of pollutants by a surface flow, values of this coefficient correspond with the tabular values found in literature. There are more factors that influence the values of the coefficient in porous media. The main one being a tortuosity of a porous medium (that is the influence of pore distortion). Accordingly, the coefficient of molecular diffusion in porous media is defined as

$$D_M = D_0 \tau, \tag{3.2}$$

where D_0 is a tabular coefficient of diffusion and τ is a coefficient of tortuosity.

The hydrodynamic dispersion is caused by a groundwater flow through a porous medium. Although the origin of the hydrodynamic dispersion is totally different from the diffusion's origin, it is still based on an equation similar to Fick's law

$$q_H = -\Theta D_H \operatorname{grad} C, \tag{3.3}$$

where q_H is a specific mass flow caused by the hydrodynamic dispersion, Θ is the soil's moisture and D_H stands for a coefficient of hydrodynamic dispersion. This approach means that both these phenomena can be superpositioned and the specific mass dispersion flow q_D is the sum of the aforementioned specific mass flows

$$q_D = q_M + q_H. \tag{3.4}$$

A coefficient of dispersion D is obtained by adding up the two previous coefficients

$$D = D_M + D_H. \tag{3.5}$$

The coefficient of hydrodynamic dispersion is often expressed as a composition of the porous velocity and of the coefficient that characterises the spreading properties of a porous medium. It is called the coefficient of dispersivity a (m)

$$D_H = a v_p. \tag{3.6}$$

This formulation stems from the idea of a one-dimensional transport. In the beginning, everything was based only on laboratory experiments that were mostly one-dimensional. Mathematical models of a transport in real conditions require a tensor form of the coefficient of dispersion (for details see Bear 1972).

$$D_{ij} = a_{ijkl}\frac{v_k v_l}{|v|} + T_{ij} D_0, \tag{3.7}$$

where a_{ijkl} is a tensor of dispersivity, v_k and v_l are components of the velocity in directions k and l, v is the length of the velocity vector and T_{ij} is the tensor of tortuosity. In many cases it is presumed that the tensor of dispersivity has only two non-zero components; a longitudinal dispersivity a_L in the direction of the velocity vector and a transversal dispersivity a_T in the direction orthogonal to the velocity vector. The coefficient of hydrodynamic dispersion for a general three-dimensional case can be expressed as (see Ackerer 1988)

$$\begin{aligned}
D_{Hxx} &= a_L \frac{v_x^2}{|v|} + a_T \left(|v|\frac{v_y^2}{v_{xy}^2} + \frac{v_z^2}{|v|}\frac{v_x^2}{v_{xy}^2}\right) & D_{Hzz} &= a_L \frac{v_z^2}{|v|} + a_T \frac{v_{xy}^2}{|v|} \\
D_{Hyy} &= a_L \frac{v_y^2}{|v|} + a_T \left(|v|\frac{v_x^2}{v_{xy}^2} + \frac{v_z^2}{|v|}\frac{v_y^2}{v_{xy}^2}\right) & & \\
D_{Hxy} &= D_{Hyx} = (a_L - a_T)\frac{v_x v_y}{|v|} & D_{Hxz} &= D_{Hzx} = (a_L - a_T)\frac{v_x v_z}{|v|} \\
D_{Hzy} &= D_{Hyz} = (a_L - a_T)\frac{v_z v_y}{|v|},
\end{aligned} \tag{3.8}$$

where v_x, v_y, v_z are velocity components in x, y, and z directions, v is the length of the vector of velocity and v_{xy} is the length of a projection of the velocity vector defined as

$$v_{xy} = \sqrt{v_x^2 + v_y^2}.$$

The coefficient of hydrodynamic dispersion depends on the geometry of a porous medium and therefore, when using models of large areas, a phenomenon often called a macrodispersion occurs. It represents the influence of the medium's non-uniformity which increases the spread of pollutants and causes a significant difference between the coefficient of dispersivity obtained in the area and that set by laboratory tests. This phenomenon has an important effect on determining the coefficient of dispersion. This is the reason for an increase in the coefficient of dispersivity with the scale of an experiment. It can be clearly seen in Fig. 3.1, where data is collected from different authors who measured values of the coefficient of dispersivity in situ (see Rahman and Hulla

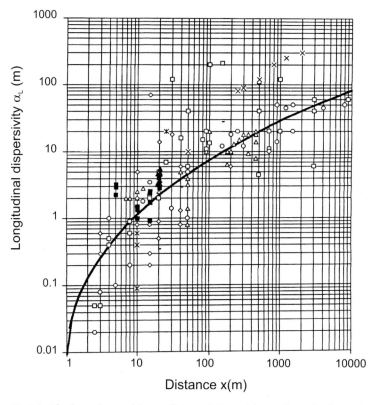

Fig. 3.1. The dependence of the coefficient of dispersivity on the scale of experiment (see Rahman and Hulla 1996)

1996). Hence, the data from laboratory experiments cannot be used in a large-scale model and we even have to be careful when using data from tracing experiments.

3.1.2
Governing Equation of Transport

The derivation of the basic equation of the transport of pollutants is based, similarly to the groundwater flow, on the conservation of mass. Let us set an infinitesimal prism in a porous medium with its orientation as in Fig. 3.2. Then one can find fluxes that move inward through the prism's face and those moving out, the latter being larger by a differential

$$\bar{q}_i = q_i + dq_i = q_i + \frac{\partial q_i}{\partial x_i} dx_i. \tag{3.9}$$

The sum of the fluxes equals a change in the concentration of a pollutant in water and can be written as

$$-\frac{\partial(\Theta C)}{\partial t} = \frac{\partial q_i}{\partial x_i}. \tag{3.10}$$

This is the simplest equation of continuity of the miscible flow. Here, we neglected any interaction of the solution with a skeleton which is the origin of adsorption and desorption phenomena (see Sect. 3.1.3) and we neglect

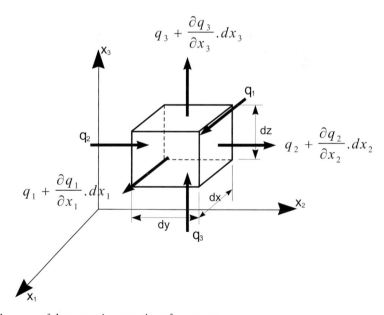

Fig. 3.2. Development of the governing equation of transport

reactions between solution's components and their spontaneous decay as well. Now we substitute a dispersion flux [see Eq. (3.4)] and an advective flux for fluxes in the equation of the transport's continuity. The advective flux is

$$q_{ci} = v_i C, \qquad (3.11)$$

where v_i, in this case, is the rate of filtration in the i-th direction. After a substitution, we have

$$\frac{\partial(\Theta C)}{\partial t} = \frac{\partial}{\partial x_i}\left(\Theta D_{ij} \frac{\partial C}{\partial x_j}\right) - \frac{\partial}{\partial x_i}(v_i C). \qquad (3.12)$$

If the transport takes place in the saturated zone, the value of moisture is constant and it is the same as the effective porosity. After derivations using the effective porosity, we obtain

$$\frac{\partial C}{\partial t} = \frac{\partial}{\partial x_i}\left(D_{ij} \frac{\partial C}{\partial x_j}\right) - \frac{\partial}{\partial x_i}(v_{Pi} C), \qquad (3.13)$$

where v_p is the porous velocity. This equation is the basic differential equation of a conservative transport of pollutants in a porous medium, that is a transport where the pollutant does not react with the environment and it does not spontaneously decay.

In some cases (mainly in a pollutants' transport through the unsaturated zone), measurements of breakthrough curves show that the front of concentration stays behind the front of moisture and that the decrease in concentration after the maximal value is slower than the increase (see Fig. 3.3). This is explained by the existence of a non-movable component of groundwater that does not take part in an advective transport. It is caused by dead-end pores and by the pellicular water (see Chap. 2.2.1). For this case, Eq. (3.13) should be derived to (see Gaudet et al. 1977)

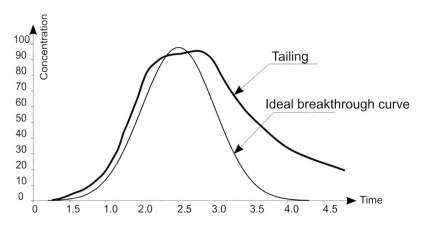

Fig. 3.3. Breakthrough curve tailing

$$\frac{\partial(\Theta_m C_m)}{\partial t} + \frac{\partial(\Theta_{im} C_{im})}{\partial t} = \frac{\partial}{\partial x_i}\left(\Theta_m D_{ij} \frac{\partial C_m}{\partial x_j}\right) - \frac{\partial}{\partial x_i}(v_i C_m). \tag{3.14}$$

The index m stands for a mobile component of groundwater and the index im for an immobile component of groundwater. An exchange between the mobile and immobile components is governed by this formula

$$\frac{\partial(\Theta_{im} C_{im})}{\partial t} = \beta(C_m - C_{im}), \tag{3.15}$$

where β is a transport coefficient (s^{-1}). This system of equations is analogous to equations describing a non-equilibrium sorption (see Sect. 3.1.5) and can be solved in a similar manner (see Chap. 6.3).

3.1.3
Sorption and Decay of Pollutants

In the previous chapter we derived the basic governing equation of the transport of pollutants when they do not react with the medium's skeleton. It is only the simplest case and in real situations these reactions can be neglected only on some occasions.

In this chapter we will focus on improving the governing equation so that it can describe some of these phenomena. Sorption means the reaction of a dissolved pollutant with soil and consists of two parts, adsorption and desorption. Adsorption is a phenomenon where the dissolved pollutant is bound to the skeleton and does not take part in the transport. Desorption works the other way round, i.e. under some circumstances, the pollutant bound to the skeleton is released back to the solution. Generally, the soil's skeleton has a negative charge. Thus, the adsorption is more apparent in solutions that include cations. With anions the adsorption has no effect and sometimes (e.g. with chlorides) it can cause a release of the (chloride) anions from the skeleton. In this case, the adsorption is negative, so a chloride solution spreads faster (for example through clay) (see Van Genuchten and Simunek 1996). To allow the equation to solve problems including adsorption, we have to add the change of pollutant mass bound to the skeleton, to Eq. (3.10).

The concentration C of a pollutant soluted in water is expressed as the mass of pollutant in a unitary volume of water (kg m^{-3}). On the contrary, the concentration S of the pollutant sorbed on the skeleton is expressed as mass of the pollutant sorbed on a unit of mass of the skeleton (the concentration is non-dimensional). The basic equation (3.10) changes to

$$-\frac{\partial(\rho S + \Theta C)}{\partial t} = \frac{\partial q_i}{\partial x_i}, \tag{3.16}$$

where Θ is the moisture and ρ is a bulk density of skeleton grains. After substituting for the discharge, as in the previous chapter, we have

$$\frac{\partial}{\partial t}(\rho S + \Theta C) = \frac{\partial}{\partial x_i}\left(\Theta D_{ij}\frac{\partial C}{\partial x_j}\right) - \frac{\partial}{\partial x_i}(v_i C) + \sum_k Z_k. \tag{3.17}$$

The last term on the right side of the equation stands for sources that were not included in either in the dispersion flux or the advective flux. This term enables us to use the equation for problems where different reactions of pollutants (such as decaying processes) occur. We assume that these Z_k expressions contain only the zero- or first-order rate terms

$$Z_k = -\mu_r \Theta C - \mu_s \rho S + \delta_r \Theta + \delta_s \rho, \tag{3.18}$$

where μ_r, μ_s are constants of a rate of decay reactions of the first order in the solution and in the sorbed part, respectively, δ_r, δ_s are constants that give a production of ions in a pollutant (kinetics of the zero order) and in the case of radioactive decay it is $\mu_r = \mu_s = \lambda$ (λ is a constant of decay), $\delta_r = \delta_s = 0$ and the formula below can be used

$$Z_k = -\lambda(\Theta C + \rho S). \tag{3.19}$$

Other degrading processes (of chemical or microbiological character) can be represented by different values of the constants above.

Before we come to the solution itself, it is essential to find the relationship between C and S. To determine this relationship, some simplifying assumptions have to be made. The solutions can be divided into two groups.

- *Solution of a equilibrium adsorption* – this is used when the velocity of the transport is a lot lower than the rate of adsorption. We suppose the adsorption is instantaneous and the rate of adsorption is considered infinite.
- *Solution of a non-equilibrium adsorption* – here the velocity of the transport is comparable with the rate of adsorption and the latter can not be neglected and has to be included in the solution.

3.1.3.1
Equilibrium Adsorption

We have already mentioned above that the rate of adsorption can be neglected. The problem of finding the relationship between C and S is thereby reduced to finding the relation between the concentrations themselves. These functions are termed isotherms even though they have nothing to do with temperature or isothermia. The simplest and most common isotherm is the linear isotherm

$$S = K_d C, \tag{3.20}$$

where K_d is called a coefficient of distribution. It is obvious that by using this relation the basic equation is simplified. After substituting it into Eq. (3.17), we have

$$(\rho K_d + \Theta)\frac{\partial C}{\partial t} = \frac{\partial}{\partial x_i}\left(\Theta D_{ij}\frac{\partial C}{\partial x_j}\right) - \frac{\partial}{\partial x_i}(v_i C) + \sum_k Z_k. \tag{3.21}$$

Equation (3.21) usually has a different form in the saturated zone where the moisture is constant and it equals the effective porosity

$$R\frac{\partial C}{\partial t} = \frac{\partial}{\partial x_i}\left(D_{ij}\frac{\partial C}{\partial x_j}\right) - \frac{\partial}{\partial x_i}(v_{pi}C) + \sum_k Z'_k, \tag{3.22}$$

where R is a retardation factor and it is given as

$$R = 1 + \frac{\rho}{m_e}K_d. \tag{3.23}$$

We altered the expressions relating to the reaction of pollutant; thus, the source term in Eq. (3.22) is denoted by Z'. This term in the saturated zone is

$$Z'_k = \left[-\mu_r + \delta_r + (\delta_s - \mu_s)\frac{\rho K_d}{m_e}\right]C. \tag{3.24}$$

For a frequent case of radioactive decay, the source term becomes

$$Z'_k = \lambda RC. \tag{3.25}$$

The linear isotherm can be easily applied in the basic governing equation, though it is only a rough approximation of the relationship between the concentrations. In most cases, the sorption is non-linear, so the retardation factor generally looks like this:

$$R = 1 + \frac{\rho}{m_e}\frac{\partial S}{\partial C}. \tag{3.26}$$

There are a number of different definitions of the non-linear formula (see e.g. van Genuchten and Šimunek 1996). The two best known isotherms are the Freundlich isotherm

$$S = K_d C^a \quad R = 1 + \frac{\rho}{m_e}aK_d C^{a-1}, \tag{3.27}$$

and the Langmuir isotherm

$$S = \frac{K_1 C}{1 + K_2 C} \quad R = 1 + \frac{\rho}{m_e}\frac{K_1}{(1 + K_2 C)^2}. \tag{3.28}$$

The basic governing equation is non-linear for both these isotherms.

3.1.3.2
Non-Equilibrium Sorption

The model of an equilibrium sorption fits the transport of pollutants in homogenous media and gives good results compared with laboratory experiments. On the other hand, it often fails when modelling a transport of strongly sorbing pollutants or organic substances (e.g. pesticides). A number of methods are used in these situations, based on the sorption's kinetics. This means that the rate of sorption is not considered infinite and, consequently, it has to be included in our equations. This type of model fits the data from laboratory experiments as well as from in-situ tracing experiments.

The simplest form is again a linear model. In literature it is often called the one-site model (see van Genuchten and Simunek 1996) and it describes the time behaviour of the sorption using a linear differential equation

$$\frac{\partial S}{\partial t} = \alpha(K_d C - S) - \mu_s S + \delta_s, \tag{3.29}$$

where α is a kinetic coefficient of the first order that characterises the rate of sorption, K_d is again the coefficient of distribution and μ_s, δ_s are constants of the zero- or first-order kinetics of the decay of a sorbed pollutant. Models using this equation are only a slight improvement compared with the traditional models that use the equilibrium sorption. The value of the coefficient α influences an often occurring phenomenon called the tailing of the breakthrough curve (see Fig. 3.3). The tailing depicts a slower decrease in the concentration in time compared with its increase. Instead of using only one governing differential equation, here we have to use a system of two equations. It is, actually, both an initial and boundary problem where the transport is governed by the first equation and the sorption is ruled by the second equation. This linear system of equations can be written as

$$\frac{\partial}{\partial t}(\rho S + \Theta C) = \frac{\partial}{\partial x_i}\left(\Theta D_{ij} \frac{\partial C}{\partial x_j}\right) - \frac{\partial}{\partial x_i}(v_i C) + \sum_k Z_k$$

$$\frac{\partial S}{\partial t} = \alpha(K_d C - S) - \mu_s S + \delta_s. \tag{3.30}$$

A further improvement of the model has two separate forms. The first involves the introduction of non-linear models of the non-equilibrium sorption, using a procedure similar to equilibrium sorption (even similar names). For example, the Freundlich model of sorption has this governing equation

$$\frac{\partial S}{\partial t} = \alpha(K_d C^a - S) - \mu_s S + \delta_s, \tag{3.31}$$

and the Langmuir model is defined as

$$\frac{\partial S}{\partial t} = \alpha\left(\frac{K_1 C}{1 + K_2 C} - S\right) - \mu_s S + \delta_s. \tag{3.32}$$

The second approach leads to a two-site model of sorption. This model is based on an assumption that particles of the soil's skeleton have two sites to which the molecules of the soluted pollutant can be bound. Molecules are bound to the first site instantaneously by equilibrium sorption and to the second one slowly by non-equilibrium sorption. The governing equations are:

$$\frac{\partial}{\partial t}[(\rho f K_d + \Theta)C] = \frac{\partial}{\partial x_i}\left(\Theta D_{ij} \frac{\partial C}{\partial x_j}\right) - \frac{\partial}{\partial x_i}(v_i C) -$$
$$- \alpha \rho[(1-f)K_d C - S] - [\Theta(\mu_r - \delta_r) + f\rho(K_d \mu_s - \delta_s)]C \qquad (3.33)$$
$$\frac{\partial S}{\partial t} = \alpha[(1-f)K_d C - S] - \mu_s S + \delta_s,$$

where f is a share of every position in the soil's skeleton. If $f = 0$, the two-site model transforms to a linear model of non-equilibrium sorption. If $f = 1$, we have a linear model of equilibrium sorption.

3.2
Transport of Immiscible Pollutants

This part of the transport of pollutants was developed mainly by solving problems connected with oil exploration. Oils are a typical group that is almost insoluble and they create an entire area in a porous medium, called a phase. Ergo, this part is often called a multiphase flow. An example can be seen in Chapter 2. It is the flow in the unsaturated zone that can be considered a flow in a two-phase system (water-air). In reality, the most frequent case of a multiphase flow is the two-phase flow (e.g. oil-water, water-air) and sometimes even a three-phase flow (e.g. oil-water-air). This kind of transport is currently present in problems of insoluble pollutants (NAPL). No pollutant is absolutely insoluble and both systems of transport (the miscible and the immiscible) are complementary to each other. However, in mathematical models we try to neglect the less important components of a phenomenon. A later chapter reviews the basic equations used in mathematical models that solve this problem. More details can be found in Bear (1972) or Barenblatt (1974).

3.2.1
General Law of Darcy

The basic characteristic of a multiple flow is the degree of saturation S_i. This is the rate of pore volume filled with the phase i. So it is obvious that

$$\sum_{i=1}^{n} S_i = 1, \qquad (3.34)$$

where n is the number of phases. Therefore, in an n-phase system there are $(n-1)$ independent degrees of saturation. In hydraulics there is capillary pressure on every line of diversion between two fluids or a fluid and a gas which is based on intermolecular forces. The interface of the phases is then formed into a large number of curved surfaces and the capillary pressure can be expressed using the curvature of the surfaces

$$p_k = \sigma\left(\frac{1}{R_1} + \frac{1}{R_2}\right), \tag{3.35}$$

where σ is a surface tension and R_1, R_2 are radii of the surfaces' curvature. These radii are comparable with the radii of pores. The pore radius can be estimated using Eq. (2.56)

$$d = \sqrt{\frac{32 k_p}{m_e}}. \tag{3.36}$$

The capillary pressure reaches tens of MPa because the coefficient of permeability is small (about $10^{-10}\,\mathrm{m}^2$). The reason for the random distribution of pores and for the irregularity in an interphase boundary is the formation of isolated particles of pollutant in the medium. These particles are so strongly bound by the capillary pressure that they are almost non-movable. The multiphase theory presumes that fluids move only in continuous areas that do not change and a distribution of every phase in pores is determined by the capillary pressure. Every phase then moves under the influence of its own pressure. Bearing this in mind, Leverett wrote Darcy's law in this new form:

$$v_{\alpha i} = -\frac{k_p k_{r\alpha}}{\mu_\alpha}\left(\frac{\partial p_\alpha}{\partial x_i} + \rho_\alpha g \frac{\partial x_3}{\partial x_i}\right) \quad i = 1,\,2,\,3 \tag{3.37}$$

where $k_{r\alpha}$ is a non-dimensional quantity that was termed a relative α-phase permeability, μ_α is a dynamic viscosity coefficient of the α-phase and p_α is pressure in the α-phase.

3.2.2
Basic Equation of Multiphase Flow

The fundament of everything below is again the continuity equation [Eq. (2.15)] which, for the α-phase, is derived into

$$\frac{\partial}{\partial t}(\rho_\alpha m_\alpha) + \frac{\partial}{\partial x_i}(\rho_\alpha v_{\alpha i}) = 0. \tag{3.38}$$

The value m_α can be expressed with the help of the rate of saturation S_α

$$m_\alpha = m S_\alpha, \tag{3.39}$$

where m is the medium's porosity. After substituting Eq. (3.39) into Eq. (3.38), we have

$$\frac{\partial}{\partial t}(m \rho_\alpha S_\alpha) + \frac{\partial}{\partial x_i}(\rho_\alpha v_{\alpha i}) = 0. \tag{3.40}$$

If we neglect the deformation of a porous medium, $m = \mathrm{const}$ and Eq. (3.40) has this new form

$$m\frac{\partial}{\partial t}(\rho_\alpha S_\alpha) + \frac{\partial}{\partial x_i}(\rho_\alpha v_{\alpha i}) = 0. \tag{3.41}$$

After neglecting the compressibility of the α-phase as well, we obtain

$$m\frac{\partial S_\alpha}{\partial t} + \frac{\partial v_{\alpha i}}{\partial x_i} = 0. \tag{3.42}$$

When we substitute for the rate of filtration $v_{\alpha i}$ [Eq.(3.36)], we acquire a governing equation of the α-phase motion

$$m\frac{\partial S_\alpha}{\partial t} - \frac{\partial}{\partial x_i}\left[\frac{k_p k_{r\alpha}}{\mu_\alpha}\left(\frac{\partial p_\alpha}{\partial x_i} + \rho_\alpha g \frac{\partial x_3}{\partial x_i}\right)\right] = 0. \tag{3.43}$$

We have these four equations for a two-phase flow with phases α and β

$$\begin{aligned}
&m\frac{\partial S_\alpha}{\partial t} - \frac{\partial}{\partial x_i}\left[\frac{k_p k_{r\alpha}}{\mu_\alpha}\left(\frac{\partial p_\alpha}{\partial x_i} + \rho_\alpha g \frac{\partial x_3}{\partial x_i}\right)\right] = 0 \\
&m\frac{\partial S_\beta}{\partial t} - \frac{\partial}{\partial x_i}\left[\frac{k_p k_{r\beta}}{\mu_\beta}\left(\frac{\partial p_\beta}{\partial x_i} + \rho_\beta g \frac{\partial x_3}{\partial x_i}\right)\right] = 0 \\
&S_\alpha + S_\beta = 1 \\
&p_\beta - p_\alpha = p_k.
\end{aligned} \tag{3.44}$$

S_α, S_β, p_α and p_β are unknown variables.

3.2.3
General Rules of Motion of Insoluble Substances in Porous Media

When this kind of substance is on the surface, gravity forces it to infiltrate into a porous medium and the rate of saturation approaches 1. Capillary forces can bring about a limited migration of the substance in a horizontal direction. The migration largely depends on the rate of the medium's anisotropy. Effects of surface tension close the substance in earth pores (see above). Some residual concentration will last in these areas because forces with their origin in the surface tension are stronger than gravity. That is true only in the case where the substance does not form a continuous phase but it is closed in the earth in isolated "drops". The retention of these substances in soil is an important process, since these drops are a secondary source of long-term pollution of groundwater. The pollution is spread by the ingress of rainwater through this area. The rainwater dissolves parts of the pollutant and transports it.

The ingress of less soluble substances decreases the quantity of mobile substances owing to the immovable drops. It can happen that the whole volume of the substance becomes divided and the motion in a vertical direction stops before reaching the groundwater level or the capillary fringe.

As soon as the pollution reaches the capillary fringe, the transport process starts to be more difficult. It depends on whether the substance has a lower or higher density compared to water (LNAPL or DNAPL). Before the pollutant reaches the fringe, it has to overcome some resistance. Hence, there must be a certain pressure created by a column of the pollutant. Pollutants lighter than water gather on the groundwater level and if their volume is large enough, they can cause a lowering of the groundwater level in places of its greatest accumulation. The DNAPL pollutants continue in their vertical motion and are only little influenced by the groundwater flow. Some non-homogeneities of a porous medium have a more decisive effect.

While the main pollutant infiltrates the medium, other processes are happening that have their own influence on the groundwater pollution, mainly the process of vaporisation of a pollutant, where new gases are created. These gases are more mobile than the liquid phase and their density often differs significantly from the soil air. They can be either soluted in the unsaturated zone or can descend to the groundwater level and be soluted there. This is the origin of the quick spread of water contamination. The process can cause even greater contamination than direct solution of the pollutant in water.

Interestingly, the degree of soil saturation by water can influence how fast pollution enters the medium. If the portion of water volume is high, the small pores are filled mainly by water and the transport of pollutants takes place only in the larger pores. If the soil is relatively dry, even the small pores become filled by a pollutant, then, however, the forming of isolated drops slows down the spread of pollution. Therefore, pollutants are generally more mobile in wet soil than in dry.

References

Ackerer Ph (1988) Random-walk method to simulate pollutant transport in alluvial aquifer or fractured rocks. In: Custodio E, et al. (eds) Groundwater flow and quality modelling. D. Reidel, Dordrecht., pp. 475–486

Barenblatt GI, Entov VM, Ryzhik VM (1974) Movement of liquids and gases in natural soils (in Russian). Nedra, Moscow

Bear J (1972) Dynamics of fluids in porous media. American Elsevier, New York

Gaudet JP, Jegat H, Vachaud G, Wierenga PJ (1977) Solute transfer with exchange between mobile and stagnant water through unsaturated sand. Soil Sci. Soc. Am. J. 4:665–671

Rahman MT, Hulla J (1996) Dispersivity and adsorption of conservative pollutants in cohesionless soils. J. Hydrol. Hydromech 44:118–131

van Genuchten MT, Simunek J (1996) Evaluation of pollutant transport in the unsaturated zone. In: Rijtema PE, Elias V (eds) Regional approaches to water pollution in the environment. Kluwer, Dordrecht., pp. 139–172

4
Weighted Residuals Method

The aim of this chapter is to explain numerical methods used to solve the basic governing equations of every case mentioned in Chapter 2. Although this book is not intended for mathematicians, it is appropriate to have a chapter explaining general basics of almost every numerical method used in mathematical models. This chapter is not written in the form of theorems and their proofs, but as a review of the area necessary to understand further problems. Anyone interested in detailed study will find it in Rektorys (1984) and the rudiments of numerical mathematics can be found in Ralston (1965) or Collatz (1964).

4.1
Basic Terms and Definitions

4.1.1
Spaces

A space is a basic term used in numerical mathematics. This section is just a review of some types of spaces. These are ordered from the most general one to the more and more special spaces where we put different requirements on elements of every space. This topic is similarly dealt with in Rektorys (1984) or Collatz (1964).

A space is generally defined as a set of elements; there can be various elements such as real numbers, complex numbers, matrices, functions and so on.

In numerical mathematics, we need to define a metrics (a distance of two elements of the space), which is done by assigning a real number $\rho(f,g)$ to these two elements f, g so that

1. $\rho(f,g) = 0$ if and only if $f = g$
2. $\rho(f,g) \leq \rho(f,h) + \rho(h,g)$ for $f, g, h \in R$

Spaces where the metrics is defined as above are called *metric spaces*. One element of the space is usually defined as a null element (notation Θ).

A space R is termed *linear* if the sum of its elements is defined. This means that the sum of elements f, g is again an element of the space R and the following relations are true

1. $\sum_{i=1}^{n} c_i f_i \in R$ where $f_i \in R$ and c_i are arbitrary real numbers.
2. $f + g = g + f$ (the commutative law).
3. $(f + g) + h = f + (g + h)$ (the associative law).

A space R is termed *normed* if it is linear and there is a number assigned to each element of the space so that:

- $\|f\| = 0$ if and only if $f = \Theta$ (a definiteness of the norm),
- $\|f + g\| \leq \|f\| + \|g\|$ (the triangle axiom),
- $\|cf\| = |c|\|f\|$ (a homogeneity of the norm).

In these definitions f, g are elements of the space R and c is any real or complex number. The normed linear space is a metric space because it is possible to define the metrics of elements f, g as a norm of their difference $\rho(f,g) = \|f - g\|$. Hereby, all requirements are met.

A *unitary* space R is a linear space where every ordered pair of elements f, g has a real or complex number $\langle f,g \rangle$ assigned to it. $\langle f,g \rangle$ is called a scalar product and has these properties

- $\langle cf,g \rangle = c \langle f,g \rangle$ where $f, g \in R$ and c is any number,
- $\langle f + g, h \rangle = \langle f,h \rangle + \langle g,h \rangle$ for $f, g, h \in R$
- $\langle f,g \rangle = \overline{\langle g,f \rangle}$, where the bar means a complex conjugate number.

The unitary space is always a normed linear space since the norm can be defined as $\|f\| = \sqrt{\langle f,f \rangle}$. To determine whether a given normed linear space is unitary, we use a theorem that says a normed linear space is unitary if and only if for any two elements of the normed linear space R the following relation is valid

$$\|f + g\|^2 + \|f - g\|^2 = 2(\|f\|^2 + \|g\|^2). \tag{4.1}$$

4.1.2 Operators

An operator is another basic term that will often be used in the following parts on numerical methods. An single-valued operator (mapping) $\mathscr{L}$ is a relation that assigns an element g of a set B of a space S to every element f of a set A of a space R. We shall write it $\mathscr{L}f = g$. The set A is called a domain of definition and the set B a range of values of the operator.

4.1.3 Function Spaces, Base Functions

In the previous section, we became acquainted with basic types of spaces. Next, we shall focus on function spaces (its elements are functions). Now, it is time to introduce the most important function spaces.

The first of them is referred to in literature as $C\langle\Omega\rangle$. It is the space of continuous real or complex functions in a closed domain Ω of a space of real numbers R_n. It is a metric space with the metrics

$$\rho(f,g) = \max_{x \in \Omega} |f(x) - g(x)|. \tag{4.2}$$

This space is a normed linear space and its norm is

$$\|f\| = \max_{x \in \Omega} |f(x)|. \tag{4.3}$$

It is not a unitary space, as it is possible to find two functions from the space which do not fulfil the theorem above.

The next important function space carries a symbol $L_2\langle\Omega\rangle$. It is a space of square-Lebesgue integrable functions. This is a normed linear space with the norm

$$\|f\| = \int_\Omega f^2(x_i) d\Omega, \tag{4.4}$$

and it is also a unitary space whose scalar product is defined as

$$\langle f, g \rangle = \int_\Omega f(x_i) g(x_i) d\Omega. \tag{4.5}$$

These functions are piecewise continuous and bounded in a closed domain Ω. Many spaces are derived from this space and they are called the Sobolev spaces. These are needed when deriving the finite element method. They have a symbol $W_2^{(n)}$, where n stands for the highest order of a derivative which is square-integrable. The norm of these spaces is

$$\|f\| = \int_\Omega \left[f^2(x_i) + \sum_{k=1}^n \left(\frac{\partial^k f}{\partial x_i^k} \right)^2 \right] d\Omega. \tag{4.6}$$

Let us look at one of the unitary spaces. Let us select a series of n functions Φ_i so that they will be linearly independent. Subsequently, we have n real constants α_i and the following relation is true only when all the constants

$$\sum_{i=1}^n \alpha_i \varphi_i = 0 \tag{4.7}$$

equal zero ($\alpha_i = 0 \ \forall i \leq n$).

A series of functions is called *total* when we can find constants α_i for any function u from the same space as the function series so that

$$\left\| u - \sum_{i=1}^N \alpha_i \varphi_i \right\| < \varepsilon, \tag{4.8}$$

where ε is any real number.

The functions of this series are often called *base functions* and the coefficients α_i bear the name of Fourier coefficients. It is suitable for the base functions to be orthonormal, that is

$$\langle \varphi_i, \varphi_j \rangle = \delta_{ij}. \tag{4.9}$$

From Eq. (4.8), it follows that the function u can be approximated by a function $u^{(N)}$ having this form

$$u^{(N)} = \sum_{i=1}^{N} \alpha_i \varphi_i. \tag{4.10}$$

A norm of the function $u^{(N)}$ is

$$\|u^{(N)}\| = \sqrt{\left\langle \sum_{i=1}^{N} \alpha_i \varphi_i, \sum_{i=1}^{N} \alpha_i \varphi_i \right\rangle} = \sqrt{\sum_{i=1}^{N} \alpha_i^2 \langle \varphi_i, \varphi_i \rangle}. \tag{4.11}$$

Every term of the sum in Eq. (4.11) is positive. Hence, the norm of $u^{(N)}$ converges to the norm of u fulfilling this condition

$$\|u^{(N)}\| \leq \|u\|.$$

4.2
Weighted Residuals Method

All mathematical modelling means "only" solving a basic governing equation of a given problem. We dealt with these equations in Chapter 2. In general, we have this operator equation

$$\mathcal{L}[f(x_i)] = p \quad x_i \in \Omega. \tag{4.12}$$

We try to find the solution in a given domain Ω which fulfils a boundary condition on a boundary Γ of the domain Ω

$$\mathcal{S}[f(x_i)] = q \quad x_i \in \Gamma. \tag{4.13}$$

The solution is in our case a function. In Chapter 2, we mentioned that, on some rare occasions, we can find an exact solution of our operator equation [function $u(\alpha_i)$]. When we substitute the solution back into Eq. (4.12)

$$\mathcal{L}(u_0) - p = 0, \tag{4.14}$$

and on the boundary Γ this relation is true

$$\mathcal{S}(u_0) - q = 0. \tag{4.15}$$

In most cases we cannot find an exact solution of Eq. (4.12). We successively try to find an approximate solution u. This is often written as a linear combination of the base functions

$$u(x_i) = \sum_{k=1}^{N} \alpha_k \Phi_k(x_i). \tag{4.16}$$

After substituting back into the initial equation, we do not obtain a value which is identically equal to zero. The equation is fulfilled except for a certain error called a residue ε

$$\mathcal{L}(u) - p = \mathcal{L}\left(\sum_{k=1}^{N} \alpha_k \Phi_k\right) - p = \varepsilon. \tag{4.17}$$

On the boundary we have

$$\mathcal{S}(u) - q = \varepsilon_\Gamma. \tag{4.18}$$

We search for an approximate solution u where the errors are minimal. The principle of the weighted residuals method demands that a weighted average of a chosen set of weight functions w_i equals zero. So the following relation must be true

$$\langle (\varepsilon - \varepsilon_\Gamma) w_j \rangle = 0 \quad j = 1 \ldots N. \tag{4.19}$$

A scalar product in a function space is defined by Eq. (4.5). Let us substitute for the residue in the relation above, whereby we obtain the basic equation of the weighted residuals method (WR method)

$$\int_\Omega \left[\mathcal{L}\left(\sum_{k=1}^{N} \alpha_k \Phi_k\right) - p \right] w_j d\Omega = \int_\Gamma \left[\mathcal{S}\left(\sum_{k=1}^{N} \alpha_k \Phi_k\right) \right] w_j d\Gamma \quad j = 1 \ldots N. \tag{4.20}$$

It can be shown (see Connor and Brebbia 1976) that this principle dominates all numerical methods for different combinations of weight functions results in different methods.

To demonstrate the mechanism of deriving some of these methods, we will show derivations of some particular methods using the WR method. We will assume for the moment that the base functions fulfil all boundary conditions exactly. Thus, the right-hand side of Eq. (4.20) equals zero. Another limiting condition is based on the definition of the scalar product. It is obvious, from previous chapters, that the weight functions have to be elements of the L^2 space to assure that the scalar product exists [see Eq. (4.5)]. The base functions have to meet higher requirements since an operator of the nth order (the order of a differential equation is identical with the highest order of a derivative in this equation) is applied on them and all such functions have to belong to the Sobolev space $W_2^{(n)}$.

4.2.1
Moments Method

Here the weight functions are exponential functions of an independent variable

$$w_j = x^{j-1}. \tag{4.21}$$

These continuous, square-integrable functions belong to the L^2 space. The basic equations can be written as

$$\int_\Omega \left[\mathscr{L}\left(\sum_{k=1}^N \alpha_k \Phi_k\right) - p \right] x_i^{j-1} d\Omega = 0 \quad j = 1\ldots N. \tag{4.22}$$

If the operator $\mathscr{L}$ is linear, Eq. (4.22) is a system of N linear equations where α_k are unknown quantities.

4.2.2
Collocation Method

This method uses weight functions in a form of the Dirac delta functions $\delta(x - x_i)$ which are defined as

$$\delta(x - x_i) = \lim_{N \to \infty} \frac{\sin[N(x - x_i)]}{\pi(x - x_i)} \tag{4.23}$$

and have one important property

$$\int_{-\infty}^{\infty} \delta(x - x_i) dx = \int_{x_i - b}^{x_i + b} \delta(x - x_i) dx = 1 \tag{4.24}$$

for any positive real number b. The weight functions are then defined as

$$w_j = \delta(x_i - x_{ij}), \tag{4.25}$$

and the basic equation has this form

$$\left[\mathscr{L}\left(\sum_{k=1}^N \alpha_k \Phi_k(x_{ij})\right) - p \right] x_{ij} d\Omega = 0 \quad j = 1\ldots N. \tag{4.26}$$

Again, in case the operator is linear, it is a system of linear equations for unknown coefficients α_k. It has to be emphasised that when calculating the α_k coefficients of the system of equations using the collocation method, there are no integrals to be calculated. The accuracy of the method depends on the choice of collocation points x_{ij}.

4.2.3
Galerkin Method

This is, these days, the most frequently used method with the same weight and base functions

$$w_j = \Phi_j. \tag{4.27}$$

The basic equation has a new form

$$\int_\Omega \left[\mathscr{L}\left(\sum_{k=1}^{N} \alpha_k \Phi_k\right) - p\right] \Phi_j d\Omega = 0 \quad j=1\ldots N. \tag{4.28}$$

If the operator is linear, system (4.28) transforms into a system of N linear equations. In contrast to previous methods, here the system of equations is symmetrical and it can be shown that the system's matrix is positively definite. This fact makes the calculation much easier by decreasing the influence of rounding errors that occur by calculation.

4.2.4
Ritz Method

This is a well-known method of minimisation of a square deviation. It can be shown that even this method can be interpreted as a weighted residuals method if the operator is linear and we use weight functions as follows

$$w_j = \mathscr{L}(\Phi_j). \tag{4.29}$$

Then we have this system of basic equations

$$\int_\Omega \left[\mathscr{L}\left(\sum_{k=1}^{N} \alpha_k \Phi_k\right) - p\right] \mathscr{L}(\Phi_j) d\Omega = 0 \quad j=1\ldots N. \tag{4.30}$$

Here is a system of N equations, similar to the Galerkin method, with a symmetric and positively definite matrix. Both these methods are identical in the case of a stationary groundwater flow where the Laplace operator is used. A disadvantage of this method is that it is difficult to apply on non-linear or more complicated operators, for which reason the Galerkin method is preferred.

4.3
Weak Solution

As mentioned above, to solve a differential equation it is necessary to choose the base functions correctly. A "classical" approach using the weighted residuals method (as shown above) requires base functions from the $W_2^{(n)}$ space. Moreover, they have to fulfil all boundary conditions. It is very difficult to meet this requirement in domains of a general shape. So we see why all methods which reduce the requirements put on base functions are considered vital. The first of such methods is a weak solution of a differential equation and the second is an inverse formulation. The weak solution is the key to the finite element method while the inverse formulation is an essential step towards the boundary element method.

Prior to starting with the weak solution of a differential equation, let us have a brief look at its boundary conditions. Until now we have presumed that the base functions, Φ_i, were chosen so that all of these conditions were fulfilled exactly.

Boundary conditions can be divided into two major groups:
- *main* (sometimes referred to as stationary or essential),
- *natural* (mobile).

The main boundary condition of a differential equation of the 2*k*th order is a condition that sets function values and values of derivatives to the (*k*-1)th order [(*k*-1)th order included] on the domain's boundary. The natural boundary condition sets derivatives of a order even higher than (*k*-1)th. The main boundary condition of the governing equation of a groundwater flow (as it is an equation of the 2nd order) sets, for example, only the function values, so it is the Dirichlet boundary condition. Other boundary conditions are natural.

Let us return to the weak solution of an operator equation. Using a classical method, we need base functions belonging to the $W_2^{(n)}$ space thus the weight functions can be only from $W_2^{(0)}$ which is actually the L_2 space. Apart from this, we demand that the base functions fulfil all boundary conditions. There are too many conditions for the base functions and only a few for the weight functions. The weak solution transfers some of the requirements made of the base functions.

In most applications, operators consist of partial derivatives, and to obtain the weak solution we use the Green theorem. This technique is best demonstrated on the Laplace differential equation of the 2nd order. The equation is usually written as

$$\Delta f = 0, \tag{4.31}$$

where the symbol Δ stands for the operator

$$\Delta = \frac{\partial^2}{\partial x_i^2}. \tag{4.32}$$

A solution by the weighted residuals is based on searching for an approximate solution u in a form defined by Eq. (4.19) which satisfies this equation

$$\int_\Omega (\Delta u) w_j d\Omega = 0 \quad j = 1 \ldots N. \tag{4.33}$$

Here, we demand that the base functions meet only main boundary conditions which are set on the Γ_1 part of the boundary. In this case, the natural boundary conditions that are given on the Γ_2 part, are supposed to be the Neumann boundary conditions

$$\frac{\partial u}{\partial \nu} = \frac{\partial u}{\partial x_i} n_i = \bar{q}, \tag{4.34}$$

where n_i is a direction cosine of an exterior normal to the Γ_2 boundary. We require that these boundary conditions are also satisfied in a weighted average

$$\int_{\Gamma_2} \left(\frac{\partial u}{\partial \nu} - \bar{q} \right) w_j d\Gamma_2 \quad j = 1 \ldots N. \tag{4.35}$$

By combining Eqs. (4.33) and (4.35), we have

$$\int_\Omega (\Delta u) w_j d\Omega = \int_{\Gamma_2} \left(\frac{\partial u}{\partial v} - \bar{q} \right) w_j d\Gamma_2 \quad j=1\ldots N. \tag{4.36}$$

The main tool to obtain the weak solution is the Green theorem (or a formula for integration by parts for functions with more variables). This theorem, assuming that functions f, g are continuous together with their derivatives of the 1st order, states that (see Rektorys 1984)

$$\int_\Omega f \frac{\partial^2 g}{\partial x_i^2} d\Omega = \int_\Gamma f \frac{\partial g}{\partial v} d\Gamma - \int_\Omega \frac{\partial f}{\partial x_i} \frac{\partial g}{\partial x_i} d\Omega. \tag{4.37}$$

Let us apply this theorem to the left side of Eq. (4.36)

$$\int_\Omega \frac{\partial u}{\partial x_i} \frac{\partial w_j}{\partial x_i} d\Omega = \int_{\Gamma_2} \bar{q} w_j d\Gamma + \int_{\Gamma_1} \frac{\partial u}{\partial v} w_j d\Gamma \quad j=1\ldots N. \tag{4.38}$$

We usually require that the base functions were chosen in order to satisfy the homogenous boundary conditions on the Γ_1 boundary. Subsequently, the last integral on the right-hand side of Eq. (4.38) vanishes. We see that the order of the u function's derivative, in the new Eq. (4.38), was lowered, and at the same time we increased the order of the weight function's derivative. Hence, the weight functions along with the base functions belong to the $W_2^{(1)}$ space. This reduction fits the Galerkin method, since it uses the same weight and base functions.

4.4
Inverse Formulation

When explaining the inverse formulation principle, we remain with the Laplace equation and its solutions. In contrast to the previous paragraph, we do not ask either the weight functions or the base functions to satisfy any boundary conditions. We will further consider two kinds of boundary conditions

- main (Dirichlet condition),
- natural [Neumann, see Eq. (4.34)].

Instead of the Green theorem shown above, we now use the Green formula. It uses two functions f, g which are continuous together with their derivatives of the 1st and the 2nd order in a domain Ω. The formula is expressed by the relation below.

$$\int_\Omega (\Delta f g - f \Delta g) d\Omega = \int_\Gamma \left(\frac{\partial f}{\partial v} g - f \frac{\partial g}{\partial v} \right) d\Gamma \tag{4.39}$$

Substituting the formula into Eq. (4.36), we gain

$$\int_\Omega u\Delta w_j d\Omega = -\int_{\Gamma_2} \bar{q} w_j d\Gamma - \int_{\Gamma_1} \frac{\partial u}{\partial v} w_j d\Gamma + \int_{\Gamma_2} u \frac{\partial w_j}{\partial v} d\Gamma + \int_{\Gamma_1} \bar{u} \frac{\partial w_j}{\partial v} d\Gamma \qquad (4.40)$$

$$j = 1 \ldots N.$$

The name inverse formulation is derived from the fact that the functions u and w "change their places" on the left-hand side of Eq. (4.40). Simultaneously, we changed the requirements put on the continuity of base functions ϕ and weight functions w. The base functions belong only to the L_2 space and the weight functions now belong to the $W_2^{(2)}$ space. As the weight functions are chosen more or less arbitrarily, we can set them to fulfil this equation

$$\Delta w = -\delta_i, \qquad (4.41)$$

where δ_i is the Dirac function [see Eq. (4.23)]. Functions that satisfy this equation are called a *fundamental solution* of the problem. If we use this fundamental solution as a weight function, the integral on the left-hand side of Eq. (4.40) can be directly replaced by a function value u_i (this follows from the properties of the Dirac function). It results in

$$-u_i = -\int_{\Gamma_2} \bar{q} w_j d\Gamma - \int_{\Gamma_1} \frac{\partial u}{\partial v} w_j d\Gamma + \int_{\Gamma_2} u \frac{\partial w_j}{\partial v} d\Gamma + \int_{\Gamma_1} \bar{u} \frac{\partial w_j}{\partial v} d\Gamma \quad j = 1 \ldots N. \qquad (4.42)$$

Now there is no integral over the domain Ω since it was replaced by a curvilinear integral along the boundary Γ. This relation is fundamental for the boundary integral method and even for the boundary element method with a specific choice of base functions Φ (see Chapter 5.3).

4.5
Numerical Methods Used in Problems of Groundwater Hydraulics

The first widespread numerical method in groundwater hydraulics (as well as in other disciplines) was the *finite differences method*. The basic idea of this method is the substitution of differences for the derivatives in all governing differential equations. The whole domain is covered by a rectangular grid and a boundary must be approximated by a staircase function. This method can be easily derived from an inflow-outflow balance in each calculation block (see Mucha and Sestakov 1987). Differential equations change into a linear system of equations. This is still used in groundwater hydraulics, in contrast to mechanics of solid bodies, where it was already abandoned in the 1960s. There are three reasons which explain this:

1. The mathematics of this method is relatively simple and requires no special knowledge of numerical mathematics.

2. The method enables the user to make a clear inflow-outflow balance in each block and to determine the balance in the whole domain.
3. The fact that you cannot match the domain's shape exactly is not such a great problem in hydrogeology because the boundary of a domain is often not known precisely.

The *finite element method* was introduced in hydrogeological modelling in the 1970s, when it had already driven the finite differences method out of mechanics. The mathematics on which it is based is more difficult, and not every non-mathematician can master it, as was the case with the previous method. The regular grid of the finite differences method was replaced by an irregular grid composed of elements of a different shape (even curved borders). Hence, the exact shape of the domain can be matched. On the other hand, one has to admit that any possibility of a transparent inflow-outflow balance has disappeared.

The theory of a new method, the *boundary element method*, developed alongside the finite element method. The theoretical basis of the method is the boundary integral method and the mathematical requirements surpass those of the finite elements method. As this method was intended to solve problems in homogenous domains, its introduction to groundwater hydraulics was slow. Everyone working in groundwater hydraulics knew that this requirement was impossible to meet. So there had to be a practical way how to deal with non-homogeneous domains before this method could have been successfully used here. Compared with other methods, this one contributed something new, its ability to match exactly the singularities connected with a point source (or a sink). The point source (sink) has been a problem in groundwater flow modelling for years. It is a well that recharges or discharges water to or from an aquifer. This source is, from a mathematical point of view, a singularity point in the governing equation. The finite differences method has to place it in the centre of a block and it has to set a yield in this point. The result is a level (or a potential) that is an average of the whole block. This level evidently does not correspond with the level measured in the well. The error decreases with increased distance from the well. An additional correction was gradually developed with the help of an additional resistance (for details see Chap. 5).

In the finite element method a source is placed in a mesh point of the grid. In the proximity of a point source the level converges to a logarithmic curve and interpolation is mostly linear or quadratic. Thus, an error occurs. The error decreases when the density of a grid increases in the well's proximity. Unfortunately, this causes little variability in changing the position of a well, which makes the setting up of a model more complicated (if we want to propose a group of wells or a hydraulic curtain).

We solve all these problems by using the boundary element method because the base functions of the planar groundwater flow are logarithmic functions, and thus exactly match the level in the proximity of a well. This well can, moreover, be placed anywhere, even outside the grid, and so it is possible to easily change the position and the number of point sources.

References

Collatz L (1964) Funktionalanalysis und numerische mathematik. Springer Berlin Heidelberg New York
Connor JJ, Brebbia CA (1976) Finite element techniques in fluid flow Butterworth, London
Mucha I, Shestakov VM (1987) Hydraulika podzemných vod. Alfa, Bratislava
Ralston A (1965) A first course in numerical analysis. McGraw-Hill, New York
Rektorys K (1984) Variační metody v inženýrských problémech a v problémech matematické fyziky. SNTL, Prague

5
Mathematical Models of Groundwater Flow

5.1
Groundwater Flow in the Saturated Zone

Groundwater flow in the saturated zone is governed by the differential equation [Eq. (2.100)] with its boundary and initial conditions that we focused on in Chap. 2. The problem is generally three-dimensional but the most common type is a planar groundwater flow (seepage under or through a dam construction and, so on). Another example is a potential planar groundwater flow towards a system of wells. We simplify both cases neglecting the 3-D character of a flow. Numerical methods can solve even 3-D problems without software and hardware difficulties, though there are some requirements connected with the use of a 3-D model (e.g. setting the coefficients of hydraulic conductivity in a vertical and a horizontal direction in every layer of the model, a bigger precision of measurement of the inflow of water to pumping wells, knowledge of the position of every layer). It is therefore advisable to use 3-D models of groundwater flow only when you have a good geological survey to gather sufficient data. There are problems that directly require the use of a 3-D model such as greate thickness of the aquifer, partially penetrated wells and partial boundary conditions.

5.1.1
Analytic Solution

A direct (analytic) solution of the governing differential equation is possible only for the domain of a simple geometric shape or for homogenous media. These were the first mathematical models, even though they were not called models at the time they were made. Nowadays, since the development of numerical methods, they are not so important, but even now there are areas suitable for an analytic solution. The most common area is the processing of pumping tests, where the majority of the methods is based on an analytic solution of a groundwater flow equation in spherical coordinates

$$S_s \frac{\partial \Phi}{\partial t} = k \frac{1}{r} \frac{\partial}{\partial r}\left(r \frac{\partial \Phi}{\partial r}\right). \tag{5.1}$$

The boundary condition of this analytic solution represents a fully penetrated well with its diameter converging to zero and with a constant yield Q. Then Eq. (5.1) with an initial condition has this form

$$\Phi = \Phi_0 - \frac{Q}{4\pi T}\int_u^\infty \frac{e^{-x}}{x}dx \quad u = \frac{r^2 S_s}{4kt}. \qquad (5.2)$$

This is a fundamental of pumping test processing. This area is large and interesting, but it is beyond the aim of this book. If you are interested in details, see (Mucha and Shestakov 1987). Another large area is analytic solutions of a stationary planar flow of groundwater (see Halek and Svec 1979). Let us assume that the flow is horizontal and that a complex plane $z = x + iy$ is identical with the horizontal plane of the flow. The flow is then characterised by a complex potential $w(z)$

$$w(z) = G(x,y) + i\psi(x,y), \qquad (5.3)$$

where $G(x,y)$ is the Girinski potential (see Chap. 2.3.5) and $\psi(x,y)$ is the flow function. We showed in Chapter 2.2.6 that both the potential and the flow function are harmonic associated functions and therefore the complex potential is an analytic function. A complex specific volume discharge $q(z)$ replaces the complex velocity and it is defined as

$$q(z) = \frac{dw}{dz} = q_x - q_y i. \qquad (5.4)$$

There can be a finite number of singular points in the domain of the flow. The singular points can be divided into different categories. The point where the specific discharge equals null ($q = 0$) is a point of stagnation. In this point these relations are true

$$\frac{\partial G}{\partial x} = 0 \quad \frac{\partial \psi}{\partial y} = 0.$$

Streamlines intersect or brake in this point. Another category includes singular points that have an infinite specific discharge ($q \to \infty$). These points are called logarithmic singularities. Examples of such singularities are sources and sinks in the region of the flow. All streamlines intersect in these points. The remaining categories include two types of singularities: a saddle point and a vortex point.

One thing that very important for the next derivations is the solution of a flow in the neighbourhood of a point source (see Halek and Svec 1979). This point source is a logarithmic singularity and the complex potential is not defined here. The potential is analytic elsewhere in the domain. The complex potential of the point source is given by this formula

$$w(z) = \frac{Q}{2\pi}\ln z + c = \frac{Q}{2\pi}\ln(re^{i\beta}) + c. \qquad (5.5)$$

where Q is the source's yield and r,b are polar coordinates of a point in the complex plane. We obtain the Girinski potential and the flow function by separating the complex potential's real and imaginary part.

$$G(x,y) + i\psi(x,y) = \frac{Q}{2\pi}\ln r + i\frac{Q}{2\pi}\beta + c$$

$$G(x,y) = \frac{Q}{2\pi}\ln r + c \tag{5.6}$$

$$\psi(x,y) = \frac{Q}{2\pi}\beta.$$

Equipotential lines are circuits with a common centre in the point source. Streamlines form a pencil of lines that go through the point source (see Fig. 5.1).

More complicated problems are solved with the help of a conform mapping composed with one of simpler cases. One example of this superposition is a point source in a groundwater flow (see Bear 1972). The complex potential is the superposition of a point source with a yield Q [Eq. (5.6)] and a simple homogenous flow in the direction of the x-axis with a discharge q_0. It can be written as

$$w(z) = q_0 z + \frac{Q}{2\pi}\ln z + c. \tag{5.7}$$

The Girinski potential and the flow function are again obtained by separating the real and the imaginary part.

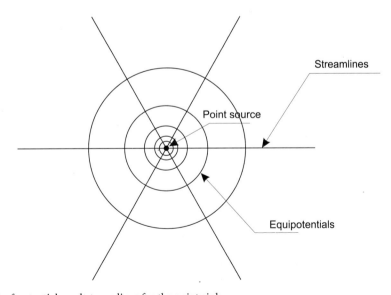

Fig. 5.1. Net of potentials and streamlines for the point sink

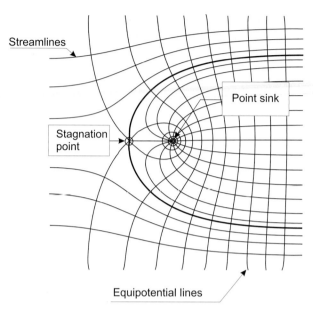

Fig. 5.2. Flow net for the single sink in uniform flow

$$G(x,y) = q_0 x + \frac{Q}{2\pi}\ln r = q_0 x + \frac{Q}{4\pi}\ln(x^2 + y^2)$$
$$\psi(x,y) = q_0 y + \frac{Q}{2\pi}\beta. \tag{5.8}$$

Discharges in the x and y direction are

$$q_x = q_0 + \frac{Q}{2\pi}\frac{x}{x^2 + y^2} \quad q_y = \frac{Q}{2\pi}\frac{y}{x^2 + y^2} \tag{5.9}$$

We can see that except for the singularity in the point source, we have a point of stagnation ($q = 0$) in a point with these coordinates

$$y = 0 \quad x = \frac{Q}{2\pi q_0}. \tag{5.10}$$

There is a hydrodynamic net for this case in Fig. 5.2.

5.1.2
Finite Differences Method

This method meant the first breakthrough in mathematical modelling in many areas, and is the oldest numerical method. Although its theoretical fundamentals were known long before, the method was fully applied only after the development of computers. This is true for all numerical methods. The mathematical principle of this method lies in replacing partial derivatives in the governing equation by differences. We can come to the same result by analysing the flow balance in a cell with finite size $\Delta x, \Delta y, \Delta z$ (see Mucha and Shestakov 1987). This method has been used for a long time and there are many detailed works on this subject (see Peaceman and Rachford 1955; Remson et al. 1970; Mucha and Shestakov 1987; Luckner and Shestakov 1976).

5.1.2.1
Basic Equations

To approximate partial derivatives by differences in a differential equation, we use the function's expansion in its Taylor series. For a function of one variable it is

$$f(x+h) = f(x) + \sum_{i=1}^{n} \frac{h^i}{i!} f^{(i)}(x) + \frac{h^{n+1}}{(n+1)!} f^{(n+1)}(\xi) \quad \xi \in (x, x+h). \tag{5.11}$$

After neglecting the terms with derivatives of high order, we have this representation of a derivative of the 1st order

$$f'(x) = \frac{1}{h}(f(x+h) - f(x)) + \frac{h}{2!} f''(\xi). \tag{5.12}$$

To represent a derivative of the 2nd order by differences, we use the function's value at $(x - h)$ and we obtain

$$f''(x) = \frac{1}{h^2}(f(x+h) - 2f(x) + f(x-h)) + \frac{h^2}{4!} f^{(4)}(\xi). \tag{5.13}$$

The last term in Eqs. (5.11) and (5.12) stands for an estimate of the error that must be counted with in this method. The best way to explain this method is to demonstrate it on an example of a solution of the governing equation of groundwater flow. The most fitting example is a one-dimensional unstationary groundwater flow. Its governing equation is

$$S \frac{\partial \Phi}{\partial t} = T \frac{\partial^2 \Phi}{\partial x^2}. \tag{5.14}$$

The storage coefficient S and the coefficient of transmissivity T are, for the sake of simplicity, both presumed constant. Let us cover the domain of the solution by a net with mesh points where we will determine the potential's values. Suppose now (again for simplicity) that the net is equidistant. This means that the distance Δx between two neighbouring mesh points is the same for any such mesh points (see Fig. 5.3). The derivative with respect to x is now replaced by a difference using Eq. (5.13). As it is an unsteady flow, the time derivative must also be replaced by a difference schema, thereby we obtain a series of time steps. We are able to determine the potential's value at a given time step if we know its value at the previous time step. The question is, which time step

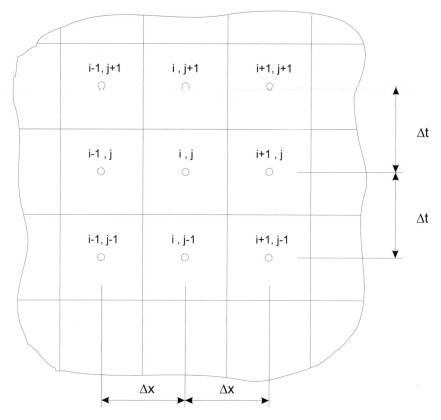

Fig. 5.3. Finite difference net

we should use to determine the value of a derivative of the 2nd order on the right side of Eq. (5.14).

To interpolate the potential between two time steps j and $j + 1$, we use a general schema

$$\Phi = (1-\xi)\Phi_j + \xi\Phi_{j+1} \quad \xi \in \langle 0,1 \rangle. \tag{5.15}$$

The ξ coefficient is used to choose among different schemes. The Crank-Nicholson schema, where $\xi = 0.5$, is the best known and the most commonly used one. Another often used schema is called the implicit schema, where $\xi = 1$. It may seem advantageous to use $\xi = 0$ (the explicit schema). Thereafter we can write

$$\frac{S}{\Delta t}(\Phi_{i,j+1} - \Phi_{i,j}) = \frac{T}{\Delta x^2}(\Phi_{i-1,j} - 2\Phi_{i,j} + \Phi_{i+1,j}). \tag{5.16}$$

The index i stands for a spatial net and the index j for a time net. We assumed above that all values in the j time step are known. The $\Phi_{i,j+1}$ value can be expressed explicitly from Eq. (5.16)

$$\Phi_{i,j+1} = \Phi_{i,j} + \frac{T\Delta t}{S\Delta x^2}(\Phi_{i-1,j} - 2\Phi_{i,j} + \Phi_{i+1,j}). \tag{5.17}$$

The other option is to use values from the same time step to express a spatial derivative (implicit schema $\xi = 1$)

$$\frac{S}{\Delta t}(\Phi_{i,j+1} - \Phi_{i,j}) = \frac{T}{\Delta x^2}(\Phi_{i-1,j+1} - 2\Phi_{i,j+1} + \Phi_{i+1,j+1}). \tag{5.18}$$

This option does not allow us to express $\Phi_{i,j+1}$ directly. We have a system of equations in every time step and values in mesh points of the spatial net are determined by solving this system of equations. Although the explicit schema does not require a solution of this system of equations, it does not enable us to use an arbitrary time step or an arbitrary size of the net's cell. To assure the stability of the solution, the following condition has to be kept

$$\frac{T\Delta t}{S\Delta x^2} < \frac{1}{2}. \tag{5.19}$$

This sets a limit to the time step. Hence, the explicit schema is only rarely used in models of groundwater flow. On the other hand, a complete implicit schema requires a solution of the system of equations in every time step. This means that a longer time is needed for the solution. In a one-dimensional case, the system has only a tridiagonal matrix, which has non-zero terms on its diagonal and on the two neighbouring diagonals of the matrix. Moreover, the matrix is symmetrical, which makes the solution much easier.

We use a similar algorithm to solve a 3-D problem. The only difference is that we must have a 3-D net. We can take an example in a planar flow with a confined groundwater level (see Chap. 2). The governing Eq. is

$$T_x \frac{\partial^2 \Phi}{\partial x^2} + T_y \frac{\partial^2 \Phi}{\partial y^2} = S \frac{\partial \Phi}{\partial t}. \tag{5.20}$$

Here, we will use a rectangular net, thus a new index will emerge. The basic relation for the implicit schema has this form

$$\frac{S_{i,j}}{\Delta t}(\Phi_{i,j,k+1} - \Phi_{i,j,k}) = \frac{T_x^{i,j}}{\Delta x^2}(\Phi_{i-1,j,k+1} - 2\Phi_{i,j,k+1} + \Phi_{i+1,j,k+1})$$
$$+ \frac{T_y^{i,j}}{\Delta y^2}(\Phi_{i,j-1,k+1} - 2\Phi_{i,j,k+1} + \Phi_{i,j+1,k+1}). \tag{5.21}$$

The indexes i, j are spatial coordinates of mesh points and the index k stands for the time schema. Obviously, the matrix is no longer tridiagonal. It will have five non-zero diagonals. Their distance from the main diagonal depends on the number of mesh points in a direction of the smaller side of the planar net's rectangle. The maximal distance of a non-zero term to the main diagonal is called a bandwidth of the system of equations. The larger the bandwidth, the larger are the requirements of the solution. As we mentioned above, this system has to be solved in every time step when using the implicit schema. Therefore, the effort of research teams focused on accelerating the solution. They introduced iteration methods such as the successive over relaxation method. These methods do not depend on the bandwidth and a so-called relaxation factor can be optimised by matrices of the finite differences method (see Kolar et al. 1972). Another method was developed by Peaceman and Rachford (1955). They proposed to split the time step into two parts. In the first part there is only a change in the x direction and in the second a change in the y direction. The whole solution is divided into two solutions of tridiagonal matrices. This is an alternating direction method (ADI method).

The basic equations are derived from the equation of continuity of a flow (see Chap. 2.1), thus every equation of the system is an equation of continuity in the neighbourhood of a mesh point. This neighbourhood is called a cell. The net in the finite differences method is orthogonal and so is the cell. In one dimension it is a line segment, a rectangle in two and a cuboid in three dimensions. It can be shown that the same set of equations as above [Eqs. (5.18) and (5.21)] can be derived from a balance of inflows and outflows in each cell. It works the other way round as well. By solving the equations we acquire the balance in every cell. This balance is very useful in further hydrogeological processing of the model's results. This is the biggest advantage of this method, keeping it very popular in hydrogeology even though it has been replaced by the finite element method in other areas (e.g. geomechanics).

5.1.2.2
Introduction of Boundary Conditions

To solve a governing differential equation, it is essential to know initial and boundary conditions. The initial conditions in the finite differences method can be given with ease. We just give initial values of the potential in each mesh point. On the other hand, the setting of the boundary conditions can lead to some problems. One problem lies in the use of an orthogonal net which this method presumes. It occurs when a boundary has a more complicated form (such as a river bed). Then the form is approximated by a homogenous net so that the mesh points (centre of the cell) are in a position where the boundary condition is given. Some authors (see e.g. Bear 1972) recommend solving this problem by interpolating values in the nearest mesh points.

When point sources are introduced, the net's mesh points are designed to match the position of point sources as precisely as they can. A yield of a point source which depends on time can be represented by a relation similar to Eq. (5.15)

$$Q = (1-\xi)Q_k + \xi Q_{k+1} \quad \xi \in \langle 0,1 \rangle. \tag{5.22}$$

The final equation of the implicit schema is

$$\frac{S_{i,j}}{\Delta t}(\Phi_{i,j,k+1} - \Phi_{i,j,k}) - Q_{i,j,k+1} = \frac{T_x^{i,j}}{\Delta x^2}(\Phi_{i-1,j,k+1} - 2\Phi_{i,j,k+1} + \Phi_{i+1,j,k+1})$$
$$+ \frac{T_y^{i,j}}{\Delta y^2}(\Phi_{i,j-1,k+1} - 2\Phi_{i,j,k+1} + \Phi_{i,j+1,k+1}), \quad (5.23)$$

where $Q_{i,j,k+1}$ is a yield in the mesh point i, j and in the $(k + 1)$th time step. The final potential is an average value of potentials in the whole cell. The real potential in the point source differs a lot from the calculated value because the commonly used differential schemes are linear. We need to thicken the net in the neighbourhood of the point source or use a different, more or less empirical, correction. The first option radically increases the number of unknown parameters together with the number of equations in the set. The second option stems from an analogue RC net where an additional resistance was added to match a radial character of a flow in the cell (see Luckner and Shestakov 1976). Now the potential's value is corrected with the help of the Dupuit formula by a difference which stands for the influence of the radial flow in the neighbourhood of the source

$$\Delta \Phi_{i,j} = \frac{Q_{i,j}}{2\pi T^{i,j}} \ln \frac{R}{r}, \quad (5.24)$$

where $Q_{i,j}$ is the source's yield, $T^{i,j}$ is the transmissivity coefficient in a given cell, r is the effective radius of the well and R is the radius of the source's zone of influence. R is the radius of a circle that has the same area as the cell containing the source

$$R = \sqrt{\frac{\Delta x_i \Delta y_j}{\pi}}. \quad (5.25)$$

$\Delta x_i, \Delta y_i$ symbolise the size of the source's cell.

5.1.3
Finite Element Method

5.1.3.1
Basic Equations

This method was introduced in the 1960s. It was first used in mechanical analysis of constructions and in the 1970s it was brought to groundwater hydraulics (see Bansky and Kovarik 1978; Kazda 1983). This area is not very typical because, compared to other areas, this method did not completely replace the previous methods (as it did in mechanics), it just opened new possibilities. We will talk about the method's development and its properties from the user's point of view in Chapter 7. Here, we focus on the mathematics of this method.

The finite element method is a weighted residuals method (see Chap. 4) and is based on the weak solution of a differential equation. We try to find an unknown function of potential Φ that is approximated, similarly to Chapter 4, by an approximate solution of this form

$$\Phi(x_l,t) = \sum_{j=1}^{n} a_j(t) N_j(x_i), \tag{5.26}$$

where N_j are base functions which are only functions of position and a_j are unknown coefficients which are functions of time. In the finite element method the base functions are specially chosen in order to be non-zero only in one domain of a simple geometric shape which is called an element. Additionally, they form an orthonormal system, which means that in one of n points in an element only one function equals 1 and the others equal zero. This point is called the node of an element. The values of a_j coefficients are values of the potential Φ in each node of an element. The Galerkin variant of this method is now almost exclusively used in the solution of a groundwater flow in the saturated zone (see Chap. 4). Here, the base functions N_j are identical with the weight functions. If we apply the same approach as in the search for the weak solution (see Chap. 4) on the basic governing equation of groundwater flow (see Chap. 2)

$$S_s \frac{\partial \Phi}{\partial t} - \frac{\partial}{\partial x_i}\left(k_{ii} \frac{\partial \Phi}{\partial x_i}\right) = 0, \tag{5.27}$$

we obtain

$$\int_\Omega \left(k_m \frac{\partial N_i}{\partial x_m}\frac{\partial N_j}{\partial x_m} a_i\right) d\Omega + \int_\Omega qN_j d\Omega - \int_\Gamma v_m k_m a_i \frac{\partial N_i}{\partial x_m} N_j d\Gamma$$
$$- \int_\Omega \left(S_s N_i N_j \frac{da_i}{dt}\right) d\Omega = 0. \tag{5.28}$$

Here, the curvilinear integral represents a natural boundary condition, v_m is a direction cosine of an exterior normal of the boundary. Equation (5.28) can be written in a matrix form

$$H\Delta + M\frac{d\Delta}{dt} + F = 0, \tag{5.29}$$

where

$$H_{ij} = \int_\Omega \left(k_m \frac{\partial N_i}{\partial x_m}\frac{\partial N_j}{\partial x_m}\right) d\Omega \quad F_i = \int_\Omega qN_j d\Omega - \int_\Gamma v_m k_m \frac{\partial N_i}{\partial x_m} N_j d\Gamma$$
$$M_{ij} = -\int_\Omega (S_s N_i N_j) d\Omega \quad \Delta_i = a_i. \tag{5.30}$$

Now, the entire domain is divided into elements of a simple shape so that the neighbouring elements are in contact either on their sides or in nodes, but do

5.1 Groundwater Flow in the Saturated Zone

not overlap. The domain integral is the sum of domain integrals over the elements. We have local matrices $\mathbf{H}^e$ and $\mathbf{M}^e$ for every element. These matrices are called the transmissivity and storage matrix of an element, respectively. We calculate the integration over every element and after adding the local matrices together, we have the final system [Eq. (5.29)]. It is a system of linear differential equations for unknown functions Δ_i which are functions of time. These are the values of the potential Φ in nodes of an element. Some numerical methods can be used to find a time dependence. An often-used one is a one-dimensional finite differences method where time is split into time steps. We assume that functions $\Delta_i(t)$ are linear in every time step and can be written as

$$\Delta = (1-\xi)\Delta_1 + \xi\Delta_2 \quad \xi \in \langle 0,1 \rangle, \tag{5.31}$$

where Δ_1 is the value at the beginning of a time step, Δ_2 is the value at the end of a time step. The coefficient ξ again serves to choose among different schemes. The best known is the Crank-Nicholson schema, which corresponds with $\xi = 1/2$. A frequently used schema is the implicit one which has $\xi = 1$.

When we substitute Eq. (5.31) into system (5.29), we obtain

$$\left(\xi\mathbf{H} + \frac{1}{\tau}\mathbf{M}\right)\Delta_2 = \left[(\xi-1)\mathbf{H} + \frac{1}{\tau}\mathbf{M}\right]\Delta_1 + (1-\xi)\mathbf{F}(t_1) + \xi\mathbf{F}(t_2). \tag{5.32}$$

This is a recurrent formula that enables us to determine the values Δ_2 at time t_2 from the values Δ_1 at time t_1. When we start from an initial condition Δ_0 at time t_0, we can obtain values in all time steps. We have to solve the system of Eqs. (5.32) in every time step and the results are the potential's values in nodes valid for the time step. If we solve the stationary flow problem, the situation is easier, because in Eq. (5.27) $S_s = 0$ and the unknown functions Δ_i are constant. Equation (5.29) changes to

$$\mathbf{H}\Delta + \mathbf{F} = \mathbf{0}. \tag{5.33}$$

This is a system of linear equations with an unknown vector Δ. In case a boundary condition of the 1st order is not defined at least in one node, the system of equations is singular.

In the finite element method different types of elements can be used. Every element is characterised by its shape and the form of its base functions. From Eq. (5.28) it follows that the base functions have to belong to the Sobolev space $W_2^{(1)}$ (see Chap. 4). The base functions are mainly polynomials. The shape of an element should stay as simple as possible in order to be able to obtain the integration over the element needed to calculate the matrices $\mathbf{H}^e$ and $\mathbf{M}^e$. The other limit to the choice of an element's shape is that there has to be a procedure (not very complicated) that covers the domain with these elements. The number of nodes in one element is governed by the order of a polynomial that is used as a base function. The number of unknown coefficients of the polynomial must be equal to the total number of Δ_i parameters to assure the problem's solvability. The number of base functions matches the number of nodes in an element.

5.1.3.2
Elements for Planar Problems

The solution of a planar problem is based either on Eq. (2.106) if it is a confined surface flow or on Eq. (2.104) if it is a flow with a phreatic surface (see Chap. 2). Basic relations of the finite element method remain similar to those derived for a 3-D case. The local matrices of an element are

$$H^e_{ij} = \int_{\Omega^e} \left(k_x B \frac{\partial N_i}{\partial x} \frac{\partial N_j}{\partial x} + k_y B \frac{\partial N_i}{\partial y} \frac{\partial N_j}{\partial y} \right) d\Omega$$

$$F^e_i = \int_\Omega qBN_j \, d\Omega - \int_{\Gamma^e} \left(v_x k_x B \frac{\partial N_i}{\partial x} N_j + v_y k_y B \frac{\partial N_i}{\partial y} N_j \right) d\Gamma \quad (5.34)$$

$$M^e_{ij} = -\int_{\Omega^e} (SN_i N_j) \, d\Omega$$

$$\Delta_i = a_i,$$

where B is the thickness of an aquifer. In the formula of the matrix $\mathbf{M}^e$, we directly used the storage coefficient S instead of the specific storage coefficient S_s. This approach is often applied in formulas of the matrices $\mathbf{H}^e$ and $\mathbf{F}^e$ where $k_m B$ is replaced by the transmissivity coefficient T_m.

The phreatic surface problem is usually non-linear, as the thickness of an aquifer is a function of the potential. The system (2.104) of an unsteady problem and the system of a steady problem are systems of non-linear equations. If the infiltration (or evaporation) from a free surface is included, the matrix $\mathbf{F}^e$ changes to

$$F^e_i = -\int_{\Gamma^e} \left(v_x k_x B \frac{\partial N_i}{\partial x} N_j + v_y k_y B \frac{\partial N_i}{\partial y} N_j \right) d\Gamma + \int_{\Omega^e} v_0 N_i \, d\Omega. \quad (5.35)$$

The simplest element of a planar case is a triangular element with three nodes in vertices of the triangle and with three base functions in the form of a linear polynomial (see Fig. 5.4). This is the first element used in groundwater flow problems (see Kovarik 1978), and it is still used because it allows an automatic preparation of the net of elements. Moreover, the net of elements can match even an irregular domain's shape and can easily be thickened. In this element the base functions are introduced by so-called planar coordinates

$$N_i = \frac{P_i}{P} = \frac{1}{2P}(a_i + b_i x + c_i y), \quad (5.36)$$

where P_i are corresponding parts of area (see Fig. 5.4) and P is the total area of the triangle. The coefficients a_i, b_i, c_i are set from a difference of coordinates in an element (see Connor and Brebbia 1976)

$$a_i = x_j y_k - x_k y_j \quad b_i = y_i - y_k \quad c_i = x_k - x_j. \quad (5.37)$$

Fig. 5.4. Linear triangular element

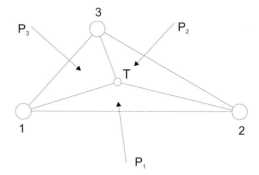

We have to integrate over the element to determine the local transmissivity and the local storage matrices of the element. Using this element, derivatives of the base functions in Eq. (5.36) are constant, thanks to the functions' linearity.

$$\frac{\partial N_i}{\partial x} = \frac{b_i}{2P} \quad \frac{\partial N_i}{\partial y} = \frac{c_i}{2P}. \tag{5.38}$$

These derivatives are used to set the transmissivity matrix of an element.

After substituting into Eq. (5.34), we can directly integrate so as to obtain this formula for the matrix of transmissivity

$$H_{ij}^e = \frac{B}{4P}(k_x b_i b_j + k_y c_i c_j). \tag{5.39}$$

A relation for the integration of polynomial functions on a triangular domain (see Connor and Brebbia 1976) can be applied on the formula of the storage matrix

$$\int_{\Omega^e} N_1^a N_2^b N_3^c d\Omega^e = \frac{a!b!c!}{(a+b+c+2)!} 2P. \tag{5.40}$$

Then we obtain the following relation for the storage matrix, and after an integration we gain

$$M_{ij}^e = \frac{1}{12} SP \quad i \neq j \quad M_{ii}^e = \frac{1}{6} SP. \tag{5.41}$$

It is obvious that this integration allows a fast calculation of the local matrices. This property, together with a simple match of a complicated shape of the domain, contributes to the popularity of this element.

Another widespread type of elements are the elements of the so-called serendipity family and these can be used in planar as well as in 3-D problems. It is a large group of elements named after a fairytale prince of Serendip. This family consists of three groups of elements:

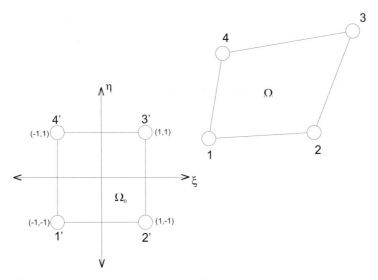

Fig. 5.5. Planar four-node isoparametric element

- isoparametric elements,
- subparametric elements,
- superparametric elements.

The isoparametric elements are the most used ones. A common feature of all these groups is that they have a more complicated shape, and to simplify the integration the element is transformed into a unitary element (see Fig. 5.5) by the polynomial transformation of coordinates below

$$x = \sum_{i=1}^{n} L_i x_i \quad y = \sum_{i=1}^{n} L_i y_i, \quad (5.42)$$

where L_i are polynomial functions (also called mapping functions), x_i and y_i are coordinates of the element's nodes. If we use the L_i functions identical with the base functions N_i, we gain the isoparametric elements. The subparametric elements use the mapping functions L_i of an order lower than the order of the base functions. The superparametric elements use the mapping functions of an order higher than the base functions.

Two types of isoparametric elements are usually used in planar problems. Both of them have a shape of a general quadrilateral. The first has one node in each vertex and four base functions.

$$N_1 = \frac{1}{4}(1-\xi)(1-\eta) \quad N_3 = \frac{1}{4}(1+\xi)(1+\eta)$$
$$N_2 = \frac{1}{4}(1+\xi)(1-\eta) \quad N_4 = \frac{1}{4}(1-\xi)(1+\eta). \quad (5.43)$$

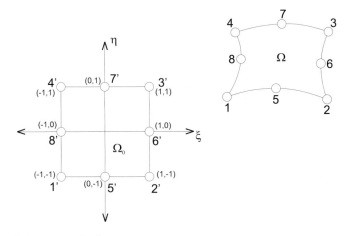

Fig. 5.6. Planar eight-node isoparametric element

The second has eight nodes, four in every vertex and four in the middle of every side. As the transformation functions are polynomials of the 2nd order, the sides of an element can be parabolic curves (see Fig. 5.6). The base functions for the vertices of the quadrilateral are

$$N_1 = \frac{1}{4}(1-\xi)(1-\eta)(-\xi-\eta-1) \quad N_2 = \frac{1}{4}(1+\xi)(1-\eta)(\xi-\eta-1)$$
$$N_3 = \frac{1}{4}(1+\xi)(1+\eta)(\xi+\eta-1) \quad N_4 = \frac{1}{4}(1-\xi)(1+\eta)(-\xi+\eta-1).$$

(5.44)

and for the middle of the sides

$$N_5 = \frac{1}{4}(1-\xi^2)(1-\eta) \quad N_6 = \frac{1}{4}(1+\xi)(1-\eta^2)$$
$$N_7 = \frac{1}{4}(1+\xi^2)(1+\eta) \quad N_8 = \frac{1}{4}(1-\xi)(1-\eta^2).$$

(5.45)

If we use elements from this family, we have to calculate the integration in a transformed system of coordinates ξ, η when deriving the local matrices. The matrices, after such a transformation, have this form

$$H_{ij}^e = \int_{-1}^{1}\int_{-1}^{1}\left(k_x B \frac{\partial N_i}{\partial x}\frac{\partial N_j}{\partial x} + k_y B \frac{\partial N_i}{\partial y}\frac{\partial N_j}{\partial y}\right)|J^e|d\xi d\eta$$

$$M_{ij}^e = \int_{-1}^{1}\int_{-1}^{1}(SN_iN_j)|J^e|d\xi d\eta,$$

(5.46)

where J^e is the Jacobian of the transformation

$$J^e = \begin{bmatrix} \dfrac{\partial x}{\partial \xi} & \dfrac{\partial y}{\partial \xi} \\ \dfrac{\partial x}{\partial \eta} & \dfrac{\partial y}{\partial \eta} \end{bmatrix}. \tag{5.47}$$

We have to transform derivatives with respect to global variables x, y to derivatives with respect to variables ξ, η in relations (5.46) using well-known formulas (see Kolář et al. 1972). Integrals in Eq. (5.46) must be calculated numerically. The Gauss integration is often used here.

Now it is time to mention a special group of elements-infinite elements. These elements originated as an attempt to eliminate a property common to the finite differences and the finite element methods. These methods can be applied only on bounded domains. There are situations in hydrogeology when some part of a boundary is very distant or not clearly defined. In these cases it would be an advantage if we could consider the part of a boundary as being at infinity. We cannot do this using ordinary methods. In the finite element method the base functions are from the $W_2^{(1)}$ space and, where no boundary condition is set, we automatically assume the existence of a natural boundary condition (natural boundary condition sets a zero inflow). We cannot obtain an unbounded domain by not setting a boundary condition but we have to use an infinite element. A few types of these elements are described in the literature. The most fitting elements are those with infinite mapping functions. They are the superparametric elements that have different base and mapping functions. The simplest one is a linear element (see Fig. 5.7) with the mapping functions

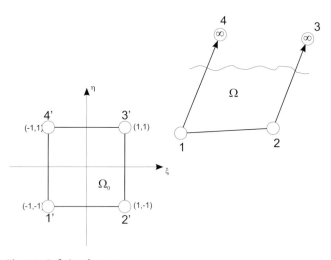

Fig. 5.7. Infinite element

$$L_1 = -(1-\xi)\frac{\eta}{1-\eta} \quad L_3 = (1+\xi)\frac{1+\eta}{2(1-\eta)}$$
$$L_2 = -(1+\xi)\frac{\eta}{1-\eta} \quad L_4 = (1-\xi)\frac{1+\eta}{2(1-\eta)},$$
(5.48)

and the base functions

$$N_1 = \frac{1}{4}(1-\xi)(1-\eta) \quad N_2 = \frac{1}{4}(1+\xi)(1-\eta).$$
(5.49)

The nodes 3 and 4 are at infinity and the nodes 1 and 2 lie on the boundary. The element is transformed to a unitary element by relation (5.42) [see Fig. (5.7)]. The matrices of transmissivity and storage are calculated similarly to the normal finite element method. We have to realise that in this special variant the potential's value at infinity equals zero. While using this method, we find out that we need an element with only one real (finite) node (e.g. in vertices of the domain). This element has this form of the mapping functions

$$L_1 = \frac{4\xi\eta}{(1-\eta)(1-\xi)} \quad L_3 = (1+\xi)\frac{1+\eta}{(1-\xi)(1-\eta)}$$
$$L_2 = -(1+\xi)\frac{2\eta}{(1-\eta)(1-\xi)} \quad L_4 = -(1+\eta)\frac{2\xi}{(1-\eta)(1-\xi)},$$
(5.50)

and only one base function

$$N_1 = \frac{1}{4}(1-\xi)(1-\eta).$$
(5.51)

Node 1 is on the boundary, the others are at infinity.

The infinite elements are not suitable in cases where we need to give a non-homogeneous boundary condition on the boundary lying at infinity. Here, some authors use quasifinite elements. These elements are mostly isoparametric with one side being a lot longer, and thereby they simulate the distance of a boundary condition. We can do this only for the isoparametric elements with quadratic base functions because it was proven that interior angles of a triangular element must be larger than 4° to reach a satisfactory precision in the solution (see Kolar et al. 1972). It was shown that the linear isoparametric elements are also unsuitable for such a use. An ideal isoparametric element, from this point of view, is the one with eight nodes and with the base and mapping functions as in Eqs. (5.44) and (5.45).

5.1.3.3
Elements for 3-D Tasks

The solution of a 3-D case has high requirements on data preparation and on processing of results as well. The large number of elements makes the preparation unclear. Therefore, 3-D models rely even more than planar ones on the

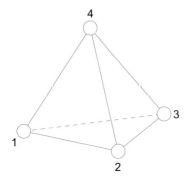

Fig. 5.8. Tetrahedron element

help of special programs that enable quick input of the domain's geometry. This is also the main concern when a shape of elements is chosen.

An analogue to the triangular element in 3-D models is a tetrahedron (see Fig. 5.8). It has again the simplest base functions in the form of a linear polynomial

$$N_i = \frac{V_i}{V} = \frac{1}{6V}(a_i + b_i x + c_i y + d_i z), \tag{5.52}$$

where V_i are volumes of the tetrahedron's parts and V is the total volume. The integration is then performed with the use of easy formulas because the function that is integrated is constant (e.g. matrix of transmissivity). Despite the properties listed above, this element is not used in 3-D models because of the intransparency of a net consisting of tetrahedrons.

There have been some attempts to introduce pentahedronal elements (see Bansky and Kovarik 1978), which should not have had this disadvantage (see Fig. 5.9). This element consists of three tetrahedrons. Integration of the local matrices is still as simple as it was in the case of a tetrahedron. This element's upper and lower surface is a triangle, which allows us to form layers consisting of elements, making the net more transparent. When constructing the net, we can use the layer structure which is common to hydrogeologic problems. Every layer is divided into triangles (upper or lower sides of an element). By the division of layers into triangles, we can use everything we know from a planar model.

Nowadays, the elements from the serendipity family are used instead. The isoparametric and the subparametric elements allow us to form layers as well. When setting up the net, we cannot forget that neighbouring elements must have a common side, edge or a vertex. Furthermore, the side and the edge must have the same shape. This is important when elements of a different shape or of a different order of a base function are combined in one model.

The simplest isoparametric element is the one of a shape of a general cuboid with line edges and sides of a shape of a line hyperboloid. It has eight nodes in its vertices (see Fig. 5.10). It is an analogue to a planar quadrilateral element. The base functions have the same form as the transformation functions

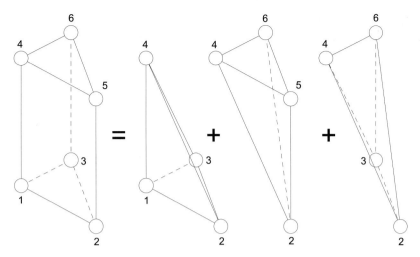

Fig. 5.9. Pentahedron element

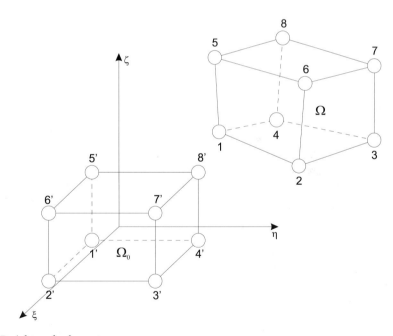

Fig. 5.10. 3-D eight-node element

$$N_1 = \frac{1}{8}(1-\xi)(1-\eta)(1-\zeta) \quad N_5 = \frac{1}{8}(1-\xi)(1-\eta)(1+\zeta)$$

$$N_2 = \frac{1}{8}(1+\xi)(1-\eta)(1-\zeta) \quad N_6 = \frac{1}{8}(1+\xi)(1-\eta)(1+\zeta)$$

$$N_3 = \frac{1}{8}(1+\xi)(1+\eta)(1-\zeta) \quad N_7 = \frac{1}{8}(1+\xi)(1+\eta)(1+\zeta)$$

$$N_4 = \frac{1}{8}(1-\xi)(1+\eta)(1-\zeta) \quad N_8 = \frac{1}{8}(1-\xi)(1+\eta)(1+\zeta).$$

(5.53)

The next isoparametric element is an element with 20 nodes (Fig. 5.11). This element has additional nodes in the middle of every edge (Kazda 1989). The base functions are embodied by N'_1 to N'_8 functions which are the base functions from Eq. (5.53). The base functions of the middles of the edges have this form

$$\begin{aligned}
N_9 &= 2N'_1(1+\eta) & N_{13} &= 2N'_5(1+\eta) & N_{17} &= 2N'_1(1+\zeta) \\
N_{10} &= 2N'_2(1-\xi) & N_{14} &= 2N'_6(1-\xi) & N_{18} &= 2N'_2(1+\zeta) \\
N_{11} &= 2N'_3(1-\eta) & N_{15} &= 2N'_7(1-\eta) & N_{19} &= 2N'_3(1+\zeta) \\
N_{12} &= 2N'_4(1+\xi) & N_{16} &= 2N'_8(1+\xi) & N_{20} &= 2N'_4(1+\zeta).
\end{aligned}$$

(5.54)

For vertices we have

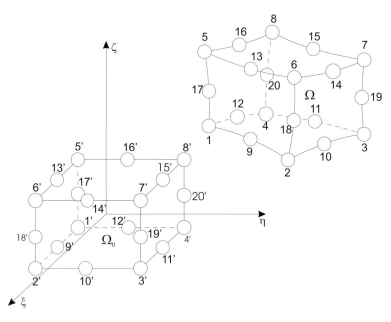

Fig. 5.11. 3-D 20-node element

$$N_1 = N_1' - \frac{1}{2}(N_{12} + N_9 + N_{17}) \quad N_2 = N_2' - \frac{1}{2}(N_9 + N_{10} + N_{18})$$

$$N_3 = N_3' - \frac{1}{2}(N_{10} + N_{11} + N_{19}) \quad N_4 = N_4' - \frac{1}{2}(N_{11} + N_{12} + N_{20})$$

$$N_5 = N_5' - \frac{1}{2}(N_{16} + N_{13} + N_{17}) \quad N_6 = N_6' - \frac{1}{2}(N_{13} + N_{14} + N_{18}) \quad (5.55)$$

$$N_7 = N_7' - \frac{1}{2}(N_{14} + N_{15} + N_{19}) \quad N_8 = N_8' - \frac{1}{2}(N_{15} + N_{16} + N_{20}).$$

In general, this element has curved sides and edges and can be used when a perfect fit of the domain's shape is required. If such precision is not needed, a subparametric element with 20 nodes will suffice. The element uses 20 base functions described by Eqs. (5.54) and (5.55). The transformation of the coordinates uses only eight functions from relation (5.53). The edges are line segments and the sides are linear hyperboloids as in the first type of an element (the one with eight nodes).

5.1.3.4
Problems Consisting of Layers

A special case occurs in systems using the finite element method, although the idea is not connected only to this method. It is called a pseudo-3-D model. The main idea lies in modelling of a system of aquifers which are considered two-dimensional and are connected between each other by one-dimensional elements that simulate an aquitard (see Fig. 5.12). We have separated the 3-D flow to a horizontal component which flows in an aquifer and to a vertical component flowing through the aquitard.

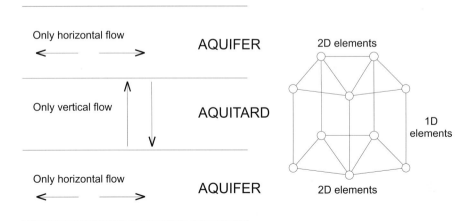

Fig. 5.12. Solution of flow through layered medium

The calculation itself is done gradually. It begins in the top aquifer and the result serves as input data for the solution of the flow in a lower aquifer. This is solved analogously, as in the first aquifer, and this continues till the lowest aquifer. It is an effective method for problems consisting of layers. It is not a full 3-D model though for this special case it is quicker than a proper 3-D model.

5.1.3.5
Setting up a Global Matrix and its Solution

In the previous paragraphs we focused on the solution of local matrices for each type of element. Now, we have to set up global matrices of transmissivity and storage from the local matrices and afterwards we should include boundary conditions in these global matrices and, last but not least, we have to solve them. A great advantage of the finite element method is the fact that neither the setting-up nor the solution depends on the type of element used, and so it does not depend on the problem to be solved.

We have already talked about the setting-up of a net: that the elements must have an edge, a side or a vertex in common and that they cannot overlap. In every node of the net we have a given number of parameters (unknown values we have to find by the solution). In groundwater flow it is mostly only one parameter-the value of the potential. This simplifies the setting-up of the global matrix because one row of the matrix corresponds with one node. Formally, the setting-up lies in the sum of the so-called extended local matrices. An extended local matrix has the same order as the global matrix and we obtain it by putting the terms of the local matrix into positions belonging to the node. The remaining terms of the new matrix equal zero. To set up the matrix we use "code numbers" of the element. A code number consists of numbers of every node in an order identical with the local numbering. The code numbers precisely define the net's topology. In older systems, these code numbers had to be given by the user, which was very complicated task and thus, of course, a source of mistakes. Presently, the systems are equipped with subprograms that generate the net automatically and create the code numbers.

We have spoken of problems with the bandwidth of a matrix in the finite differences method. The same problem emerges here as well. The price we have to pay for a net matching the domain's shape is a wider bandwidth, which means a longer time is required for the solution. This bandwidth is defined as the maximal difference of node numbers in one element. Generally, the time of a solution of a linear system is in a linear relationship with the number of equations but in a quadratic relationship with the bandwidth. Ergo, the numbering of the net was from the beginning of great importance. It is obvious that optimising the numbering of the net is the only way to save time. Every modern system optimises the numbering automatically. It is interesting that you do not have to minimise the bandwidth directly, since some methods of solution of linear systems use a different optimisation. For example in the *frontal solution* used with isoparametric elements (see Irons 1970) the time depends more on the numbering of the elements than of the nodes. Another example is the *skyline method*, where the envelope of the matrix is the decisive parameter and not the maximal bandwidth.

5.1.3.6
Introduction of Boundary Conditions

One row in the matrix of the system of equations stands for the equation of continuity in one node. If it is an inner node where there is no source, the right side of the equation equals zero. If there is a point source in this node (see Chap. 2), there is a yield of this source on the right side. The boundary condition of the 2nd kind (the Neumann condition) is introduced by determining the quantity q_0 flowing towards a boundary node from areas of elements that meet in this node (see Fig. 5.13). This area is called the zone of influence of this node.

The value of the potential in a boundary node is a boundary condition of the 1st kind. This condition can be introduced in several ways. The most correct one (from a mathematical point of view) is the lowering of an order of the system of equations by eliminating equations using the known values (the condition gives these values). This approach requires a lot of work, thus another method is used (namely a modification of the diagonal term of the matrix). This modification means that the diagonal term and the potential value on the right side are both multiplied by the same large factor. This way is quick and easy, but it increases the error of the method used for the solution of the system.

The aforementioned introduction of the boundary condition of the 2nd kind can only be done if the elements use the Lagrangian polynomials as base

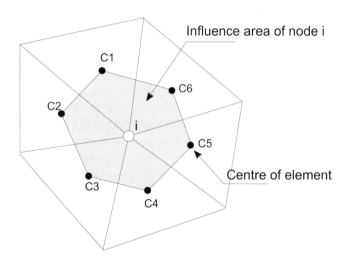

Fig. 5.13. Influential area of node

functions. All types of elements in this chapter satisfy this condition. With elements based on the Hermite interpolation polynomial, not only the value of the potential but also the values of its derivatives are unknown. In this case, the Neumann condition is introduced in the same manner as the boundary condition of the 1st kind by giving the values of all unknown quantities.

Boundary conditions of the 3rd kind are transformed to the 2nd kind by determining the inflow to the domain from the estimated value of the potential in the node with a condition of the 3rd kind. It is then corrected by iterations.

The inclusion of point sources into the solution may seem an easy task. We only need to have the node in the position of the point source and we have to put the source's yield on the right side of an equation that belongs to the node. We must not forget that groundwater level is logarithmic in the neighbourhood of a point source and the finite element method approximates parts of it by polynomials. It is evident that the error is greater when the net in the source's neighbourhood is not so dense. The problem is solved by increasing the density of elements in the neighbourhood of a point source. This increases the number of unknown parameters as well as the bandwidth of the final system of equations. The consequence thereof is that, with every change in a point source's position, we have to change the whole net. We can use the approach as in Chap. 5.1.2.2 and determine an increment of the potential as in Eq. (5.24). In this case, we have to transform the node's zone of influence to a circle with the same area (or to a sphere with the same volume in 3-D).

5.1.4
The Boundary Element Method

This method was developed in parallel with the finite element method. We can even find traces that prove that the idea is older, its predecessor being the boundary integrals method. It is based on the inverse formulation discussed in Chap. 4. Its theoretical basis can be found in the work of Jaswon (1963). The principles of the finite element method were added to the boundary integrals method in the 1980s and this revised method was called the boundary element method. It was brought to hydraulics a few years later (see Kovarik et al. 1987). The main reason for the delay was that the method was aimed to solve problems in homogenous domains and it presents even greater difficulties than the aforementioned methods when coping with the non-homogeneities which are so characteristic of the groundwater hydraulics. Despite the complications, this method is successfully used in hydraulics.

5.1.4.1
Planar Problem Solution

We will demonstrate the method on the example of the steady groundwater flow with a confined surface. This flow is governed by a differential equation of this form (see also Chap. 2)

$$T_x \frac{\partial^2 \Phi}{\partial x^2} + T_y \frac{\partial^2 \Phi}{\partial y^2} = 0 \tag{5.56}$$

with these boundary conditions:

- main boundary condition (see also Chap. 4) $\Phi = \bar{u}$ on the Γ_1 boundary,
- natural boundary condition $q = \bar{q}$ on the Γ_2 boundary, where q stands for a flux through the boundary and so we can write

$$q = T_x v_x \frac{\partial \Phi}{\partial x} + T_y v_y \frac{\partial \Phi}{\partial y}.$$

Equation (5.56) is transformed by the following transformation of coordinates

$$\tilde{x} = x \quad \tilde{y} = y \sqrt{\frac{T_x}{T_y}} \tag{5.57}$$

to the Laplace equation

$$\frac{\partial^2 \Phi}{\partial \tilde{x}^2} + \frac{\partial^2 \Phi}{\partial \tilde{y}^2} = 0. \tag{5.58}$$

To solve this equation we use the weighted residuals method (see Chap. 4.2) and we look for an approximate solution as follows

$$u(x,y) = \sum_{i=1}^{n} a_i N_i(x,y), \tag{5.59}$$

where N_i are the base functions. After substituting the approximate solution to the Eq. (5.58) and to the boundary conditions, we have non-zero residuals

$\varepsilon = \Delta u \neq 0$ in domain Ω,

$\varepsilon_1 = u - \bar{u} \neq 0$ on boundary Γ_1,

$\varepsilon_2 = q - \bar{q} \neq 0$ on boundary Γ_2.

The weighted residuals method requires that the weighted average equals zero. This means

$$\int_\Omega (\Delta u) w_k d\Omega = \int_{\Gamma_2} (q - \bar{q}) w_k d\Gamma_2 - \int_{\Gamma_1} (u - \bar{u}) \frac{\partial w_k}{\partial v} d\Gamma_1, \tag{5.60}$$

where w_k are weight functions and $\dfrac{\partial w_k}{\partial v}$ are derivatives of weight functions in a direction of the exterior normal.

Until now we have taken the same steps as in the finite element method but, we did not assume that the approximate solution fulfils the main boundary condition. Subsequently, we use the inverse formulation instead of the weak solution (see Chap. 4.4) and we obtain

$$\int_\Omega (\Delta w_k) u d\Omega = \int_{\Gamma_2} u \frac{\partial w_k}{\partial v} d\Gamma_2 + \int_{\Gamma_1} \bar{u} \frac{\partial w_k}{\partial v} d\Gamma_1 - \int_{\Gamma_1} q w_k d\Gamma_1 - \int_{\Gamma_2} \bar{q} w_k d\Gamma_2. \tag{5.61}$$

The fundamental of the boundary element method is the use of weight functions in the form of a fundamental solution of the basic equation below:

$$\Delta w_k + \delta_k = 0, \tag{5.62}$$

where δ_k are the Dirac delta functions. These functions are infinite at $x = x_k$ and in the remaining points the function's value is zero. Additionally, each function satisfies the following condition

$$\int_\Omega \delta_k d\Omega = 1. \tag{5.63}$$

Let us substitute the results of Eq. (5.62) for Δw_k in Eq. (5.61) and thereby on the left side we gain

$$\int_\Omega (\Delta w_k) u d\Omega = \int_\Omega (-\delta_k) u d\Omega = -u_k. \tag{5.64}$$

The whole equation has a new form

$$u_k + \int_{\Gamma_2} u \frac{\partial w_k}{\partial v} d\Gamma_2 + \int_{\Gamma_1} \bar{u} \frac{\partial w_k}{\partial v} d\Gamma_1 = \int_{\Gamma_1} q w_k d\Gamma_1 + \int_{\Gamma_2} \bar{q} w_k d\Gamma_2. \tag{5.65}$$

This choice of weight functions eliminates any need for integration over the domain and all integrals that remained are integrals along the domain's boundary. The fundamental solution in the case of a flow defined by Eq. (5.62) is a function of a point source (see Halek and Svec 1979) which in a 2-D case is

$$w_{ik} = \frac{1}{2\pi} \ln\left(\frac{1}{r_{ik}}\right), \tag{5.66}$$

where r_{ik} is a distance between the points i and k.

For the sake of simplicity we make no difference between the unknown values u, q and the given values. Hence, the Eq. (5.65) becomes

$$u_k + \int_\Gamma u \frac{\partial w_k}{\partial v} d\Gamma = \int_\Gamma q w_k d\Gamma. \tag{5.67}$$

This equation is true in case the point is inside the domain Ω. As the nodes in the boundary element method are usually on the boundary of a domain, we should investigate how Eq. (5.67) changes when the point lies on the boundary (see Brebbia and Walker 1980). In order to do this, we define a neighbourhood Γ^ε of the point as a circle with a radius ε (see Fig. 5.14). We presume that the point k lies on the boundary Γ of a domain and in its neighbourhood the boundary is modified so as to include a semicircle neighbourhood Γ^ε with its radius converging to zero. The boundary is assumed to be smooth. Let us investigate a passage to limit on the right side of Eq. (5.67)

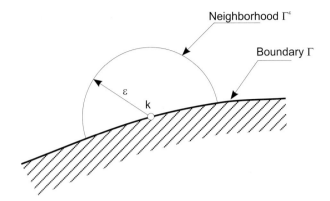

Fig. 5.14. Boundary point with semicircle neighbourhood

$$\lim_{\varepsilon \to 0}\left\{\int_{\Gamma^\varepsilon} qw_k d\Gamma\right\} = \frac{1}{2\pi}\lim_{\varepsilon \to 0}\left\{\int_{\Gamma^\varepsilon} q\ln\left(\frac{1}{\varepsilon}\right)d\Gamma\right\} = -\frac{1}{2}\lim_{\varepsilon \to 0}\{q\varepsilon\ln\varepsilon\} \equiv 0. \quad (5.68)$$

We see that this limit converges to zero. This implies that the integral on the right side of Eq. (5.67) is continuous even when the point k lies on the boundary. The integral on the left side behaves differently

$$\lim_{\varepsilon \to 0}\left\{\int_{\Gamma^\varepsilon} u\frac{\partial w_k}{\partial v} d\Gamma\right\} = -\frac{1}{2\pi}\lim_{\varepsilon \to 0}\left\{\int_{\Gamma^\varepsilon} u\frac{1}{\varepsilon} d\Gamma\right\} = -\frac{1}{2}\lim_{\varepsilon \to 0}\{u\} = -\frac{1}{2}u_k. \quad (5.69)$$

This means that when the point k lies on the boundary, the integral on the left side of Eq. (5.67) is discontinuous. Equation (5.67) for the point k on the boundary is derived to

$$c_k u_k + \int_\Gamma u\frac{\partial w_k}{\partial v} d\Gamma = \int_\Gamma qw_k d\Gamma \quad (5.70)$$

and this is the basic equation of the boundary element method. The coefficient c_k equals 0.5 for a smooth boundary and is different for a boundary with a jump discontinuity.

Now we use methods similar to FEM with the difference that we divide only the boundary of a domain into elements. The inverse formulation decreases the dimension of a problem so we use 1-D elements in a 2-D problem. As mentioned in Chap. 4.4, we changed the requirements put on every function. The weight functions have to be from the $W_2^{(2)}$ space and the base functions from the $W_2^{(0)}$ space (that is the L_2 space). The weight functions are now the fundamental solution and so the conditions put on them are met. The base functions on the other hand, are given by the type of an element and they can be, in contrast to FEM, polynomials of an even lower order. The potential u and the flux q in the direction of an exterior normal are both approximated with the help of the base functions

$$u(x,y) = \sum_{i=1}^{n} u_i N_i(x,y) \quad q(x,y) = \sum_{i=1}^{n} q_i N_i(x,y), \tag{5.71}$$

where u_i and q_i are potential and flux values in the nodes of an element. When we divide the boundary Γ into N elements, Eq. (5.70) can be written as

$$c_k u_k + \sum_{j=1}^{N} \int_{\Gamma_j} u \frac{\partial w_{kj}}{\partial v} d\Gamma = \sum_{j=1}^{N} \int_{\Gamma_j} q w_{kj} d\Gamma. \tag{5.72}$$

The simplest element is the one with a constant potential and flux with only one node in the middle of the element. Then $N_1 = 1$ and $n = 1$ thus Eq. (5.72) is easily derived to

$$c_k u_k + \sum_{j=1}^{N} u_j \tilde{H}_{kj} = \sum_{j=1}^{N} q_j G_{kj}, \tag{5.73}$$

where

$$\tilde{H}_{kj} = \int_{\Gamma_j} \frac{\partial w_{kj}}{\partial v} d\Gamma_j \quad G_{kj} = \int_{\Gamma_j} w_{kj} d\Gamma_j. \tag{5.74}$$

After introducing a new symbol

$$H_{kj} = \tilde{H}_{kj} + c_k \delta_{kj} \tag{5.75}$$

we can write Eq. (5.73) in a matrix form

Hu = Gq. (5.76)

If we realise that we know the value u on the Γ_1 boundary and the value q on the Γ_2 boundary, we can put the matrices **H** and **G** into one system of equations and the solution of that system yields the remaining unknown values of the potential and the flux in elements on the boundary. The potential's value in every point inside the domain Ω can be determined from the known values on the boundary using Eq. (5.67). In this case, the coefficient c_k equals 1. To determine the terms of the matrices **H** and **G**, we substitute relation (5.66) for the weight functions. Equation (5.74) is then

$$G_{ij} = \frac{1}{2\pi} \int_{\Gamma_j} \ln\left(\frac{1}{r_{ij}}\right) d\Gamma_j$$

$$\tilde{H}_{ij} = -\frac{1}{2\pi} \int_{\Gamma_j} \frac{D}{r_{ij}^2} d\Gamma_j, \tag{5.77}$$

where r_{ij} is the distance between the points i and j, D is the perpendicular distance from the point i to the element j (see Fig. 5.15). In the case of setting up matrices to calculate the u, q values in boundary elements, the point i is always the node of the element where we set up the equation. The points j lie on the

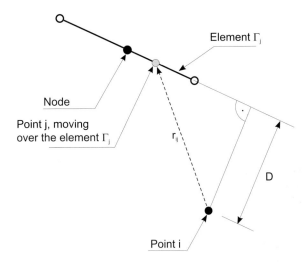

Fig. 5.15. Constant boundary element

element that is integrated over. Integrals can be calculated by numerical methods (see Brebbia and Walker 1980) or analytically in case the base functions have a form of a constant polynomial (see Broz and Prochazka 1987).

5.1.4.2
Complicated Planar Elements

Apart from constant elements, we can obviously use elements with base functions of a higher order. We can use, for example, an element where u, q functions are linear. The linear element has two nodes and we look for unknown values of the functions u, q in these points. The base functions are

$$N_1 = \frac{1}{2}(1-\xi) \quad N_2 = \frac{1}{2}(1+\xi). \tag{5.78}$$

The curved part of the boundary can be divided into elements whose base functions are quadratic polynomials. These elements have three base functions

$$N_1 = \frac{1}{2}\xi(\xi-1) \quad N_2 = (1-\xi)(1+\xi) \quad N_3 = \frac{1}{2}\xi(\xi+1) \tag{5.79}$$

and three nodes where we determine the flux q and the potential u. Here, we have to introduce local matrices $\mathbf{H}^e$ and $\mathbf{G}^e$ similarly to the FEM. The basic Eq. (5.72) changes to

$$c_i u_i + \sum_{j=1}^{N} \mathbf{u}_j^e \mathbf{H}_j^e = \sum_{j=1}^{N} \mathbf{q}_j^e \mathbf{G}_j^e, \tag{5.80}$$

where the terms of matrices $\mathbf{H}^e$ and $\mathbf{G}^e$ are

$$H^e_{kj} = \int_{\Gamma_j} N_k \frac{\partial w_{ij}}{\partial v} d\Gamma_j \quad G^e_{kj} = \int_{\Gamma_j} N_k w_{ij} d\Gamma_j. \tag{5.81}$$

Hitherto, this was a standard approach in FEM. There is a slight problem, however. In the boundary element method problems occur in vertices of the domain (see Fig. 5.16). The usual assumption is that elements that have a common node have the same value in this node. In a vertex this can be true for the potential u but it can not be true for the flux q, which has a different value and direction in both elements. This is a serious complication that can be coped with in two different ways.

The first option doubles the common node so we have two independent nodes with the same position and we can set the same potential value and different values of the flux in the nodes. The direction of the flux is the same as that of the exterior normal to the domain's border.

The second option is that we have two fluxes in one node – the first from the right and the second from the left. Then there are three unknown quantities in one node – a potential, a right flux and a left flux. When giving the boundary conditions, we have to assure that two of them are given and we look for the third. This increases the number of equations and requirements of the solution. Therefore, some authors developed a *hybrid* (or combined) element (see Martin et al. 1980), which is a combination of a constant and a linear element. The potential is approximated by the linear base functions (5.78) and the flux is constant in each element. The potential is given in nodes in the vertices of an element and the discharge is given in the middle of an element. This solves even the problem with vertices. This element is advantageous if we combine the boundary element method with the finite element method because it resembles an element from the FEM. In the finite element the flux is always approximated by a polynomial of an order lower by one than the order of the potential's polynomial (flux is a potential's derivative).

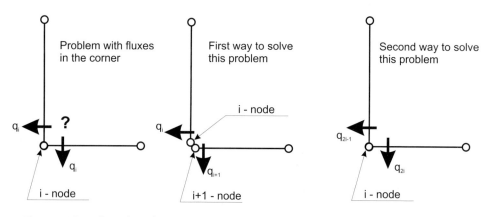

Fig. 5.16. Linear boundary element

5.1.4.3
Problem of Flow with Free-Surface and Planar Inflow

We often come across a situation where water flows into (out of) the domain in the form of a planar inflow (or outflow). A typical example is precipitation or evaporation. This phenomenon is closely connected with free-surface flow. We derived the governing equation of such flow in an unsteady case in Chap. 2. A steady case is governed by

$$\frac{\partial}{\partial x}\left(k_x B \frac{\partial h}{\partial x}\right) + \frac{\partial}{\partial y}\left(k_y B \frac{\partial h}{\partial y}\right) + v_0 = 0, \tag{5.82}$$

where B is the thickness of an aquifer. Let us remember (see Chap. 2) that the pressure on the surface is the atmospheric pressure and so $\Phi = h$. This equation is non-linear because the thickness of an aquifer is a function of the groundwater level. The linearisation of this problem is based on the assumption that the underlying stratum is horizontal ($B = h$). The next two relations are valid

$$\frac{\partial}{\partial x}\left(k_x h \frac{\partial h}{\partial x}\right) = \frac{k_x}{2}\frac{\partial^2(h^2)}{\partial x^2} \quad \frac{\partial}{\partial y}\left(k_y h \frac{\partial h}{\partial y}\right) = \frac{k_y}{2}\frac{\partial^2(h^2)}{\partial y^2}. \tag{5.83}$$

Equation (5.82) can have this form

$$k_x \frac{\partial^2(h^2)}{\partial x^2} + k_y \frac{\partial^2(h^2)}{\partial y^2} + 2v_0 = 0. \tag{5.84}$$

After a substitution of $h' = h^2$, we have

$$k_x \frac{\partial^2 h'}{\partial x^2} + k_y \frac{\partial^2 h'}{\partial y^2} + 2v_0 = 0. \tag{5.85}$$

Here, this coordinates' transformation is used

$$\tilde{x} = x \quad \tilde{y} = y\sqrt{\frac{k_x}{k_y}}, \tag{5.86}$$

and it results in the Poisson equation

$$\frac{\partial^2 h'}{\partial \tilde{x}^2} + \frac{\partial^2 h'}{\partial \tilde{y}^2} + \frac{2v_0}{k_x} = 0. \tag{5.87}$$

We use an approach analogous to the one used in the flow with a confined surface. An approximate solution u is presumed in the form given by Eq. (5.59). The inverse formulation, which is a fundamental of the boundary element method, is now used in a new form

$$\int_\Omega (\Delta w_k) u d\Omega + \int_\Omega \frac{2v_0}{k_x} w_k d\Omega = \int_{\Gamma_2} u \frac{\partial w_k}{\partial v} d\Gamma_2 + \int_{\Gamma_1} \bar{u} \frac{\partial w_k}{\partial v} d\Gamma_1 \\ - \int_{\Gamma_1} q w_k d\Gamma_1 - \int_{\Gamma_2} \bar{q} w_k d\Gamma_2. \quad (5.88)$$

The weight functions are again a fundamental solution and the basic equation of this method is

$$c_k u_k + \int_\Gamma u \frac{\partial w_k}{\partial v} d\Gamma - \int_\Omega \frac{2v_0}{k_x} w_k d\Omega = \int_\Gamma q w_k d\Gamma. \quad (5.89)$$

Although Eq. (5.89) is similar to Eq. (5.70), there is a major change. There is an integral over the domain that does not fit the boundary character of the method. To solve the case of a planar inflow, we have to integrate over the domain, which is done by numerical methods. First, we divide the domain into subdomains of a simple form where the integration can be done. The subdomains are not called elements on purpose to avoid a mistake. They can be larger than the FEM elements and there no such requirements as in FEM are put on their shape and position.

5.1.4.4
Solution of a Non-Homogenous Domain

All relations we have derived in the boundary element method assume that the coefficient of hydraulic conductivity (or the coefficient of transmissivity) is the same in the entire domain (a homogenous domain). It is obvious that this assumption is never fulfilled in the porous media. The domain must be split into zones to allow the solution of a non-homogenous domain by the boundary element method. Every zone has its matrices H and G. Apart from the boundary elements, we have elements that separate the zones-interzone elements (see Fig. 5.17). The boundary elements have their boundary condition given and the interzone must satisfy two conditions – a condition of continuity and a condition of compatibility. The former sets the potential's value in nodes of the interzone element to be the same for both zones that meet here

$$u^A = u^B. \quad (5.90)$$

A, B are indexes of each zone.

The latter requires that the outflow from the zone through an interzone element is the same as the inflow into the other zone through the same element

$$q^A = -q^B. \quad (5.91)$$

In this case we should transform the coordinates and derive this condition with respect to the coefficients of transmissivity (or hydraulic conductivity) in the neighbouring zones (for more information see Kovarik 1993). The basic equation of this method [Eq. (5.76)] can be written for each zone as

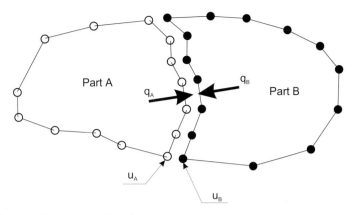

Fig. 5.17. Solution of an non-homogenous domain

$$[\mathbf{H}_O^A, \mathbf{H}_I^A]\begin{bmatrix}\mathbf{u}_O^A\\\mathbf{u}_I^A\end{bmatrix}=[\mathbf{G}_O^A, \mathbf{G}_I^A]\begin{bmatrix}\mathbf{q}_O^A\\\mathbf{q}_I^A\end{bmatrix}\quad [\mathbf{H}_O^B, \mathbf{H}_I^B]\begin{bmatrix}\mathbf{u}_O^B\\\mathbf{u}_I^B\end{bmatrix}=[\mathbf{G}_O^B, \mathbf{G}_I^B]\begin{bmatrix}\mathbf{q}_O^B\\\mathbf{q}_I^B\end{bmatrix}. \qquad (5.92)$$

The index O stands for nodes on the boundary of a domain and the index I for the interzone nodes. After applying the conditions of continuity and compatibility [Eqs. (5.90) and (5.91)], we obtain

$$\begin{bmatrix}\mathbf{H}_O^A & \mathbf{H}_I^A & 0\\ 0 & \mathbf{H}_I^B & \mathbf{H}_O^B\end{bmatrix}\begin{bmatrix}\mathbf{u}_O^A\\\mathbf{u}_I\\\mathbf{u}_O^B\end{bmatrix}=\begin{bmatrix}\mathbf{G}_O^A & \mathbf{G}_I^A & 0\\ 0 & -\mathbf{G}_I^B & \mathbf{G}_O^B\end{bmatrix}\begin{bmatrix}\mathbf{q}_O^A\\\mathbf{q}_I\\\mathbf{q}_O^B\end{bmatrix}. \qquad (5.93)$$

The system must be rearranged depending on which boundary conditions are given in the boundary elements. The unknown parameters in the interzone nodes are the values $\mathbf{u}_I$ and $\mathbf{q}_I$ which are determined from system (5.93). We have the same number of equations for nodes in an interzone element as is the number of zones that have this element in common. A rather simpler situation occurs when u, q are constant because there are only two zones that share every interzone node.

5.1.4.5
Solution of 3-D Problems

The basic fundamentals of the method remain untouched in the solution of a 3-D problem. It is again necessary to split the domain into homogenous parts (zones) and to divide the boundaries of the zones into elements. In contrast to previous cases, these elements are planar and have a simple shape. The simplest element is a triangular element and the base functions are from the L_2 space. It is an element with constant base functions and they have the same value for each element; the value equals 1. This element has only one node in its centre.

This all stems from the basic equation of the steady flow with a confined surface

$$k_x \frac{\partial^2 \Phi}{\partial x^2} + k_y \frac{\partial^2 \Phi}{\partial y^2} + k_z \frac{\partial^2 \Phi}{\partial z^2} = 0 \tag{5.94}$$

with main and natural boundary conditions on Γ_1 and Γ_2 parts of the boundary, respectively. The flux q through the boundary is defined as

$$q = k_x v_x \frac{\partial \Phi}{\partial x} + k_y v_y \frac{\partial \Phi}{\partial y} + k_z v_z \frac{\partial \Phi}{\partial z}. \tag{5.95}$$

This equation is derived by a coordinate transformation

$$\tilde{x} = x \quad \tilde{y} = y\sqrt{\frac{k_x}{k_y}} \quad \tilde{z} = z\sqrt{\frac{k_x}{k_z}} \tag{5.96}$$

into the Laplace equation

$$\frac{\partial^2 \Phi}{\partial \tilde{x}^2} + \frac{\partial^2 \Phi}{\partial \tilde{y}^2} + \frac{\partial^2 \Phi}{\partial \tilde{z}^2} = 0. \tag{5.97}$$

Deriving the basic equation of the method's stays formally the same as in Section 5.1.4.1. The weight function has obviously changed, but it is still a fundamental solution of the problem. It is a function of a point source in 3-D, which has this form

$$w_{kj} = -\frac{1}{4\pi r_{kj}}. \tag{5.98}$$

This choice of the weight function eliminates the integral over the domain and we acquire this basic equation

$$\mathbf{H}\mathbf{u} = \mathbf{G}\mathbf{q}. \tag{5.99}$$

This is an exact form of Eq. (5.76) for a planar case. The differences are in the manner of calculation of the matrices' terms. The terms of matrices $\mathbf{H}$ and $\mathbf{G}$ are calculated by integrals over the surface of the element Γ. To calculate the integral we use a transformation into a plane of the element (into coordinates ξ_1, ξ_2; see Fig. 5.18).

The next relation is true for the transformation (see Brebbia et al. 1984)

$$d\Gamma = \left| \frac{\partial \vec{R}}{\partial \xi_1} \times \frac{\partial \vec{R}}{\partial \xi_2} \right| d\xi_1 d\xi_2 = |\vec{G}| d\xi_1 d\xi_2. \tag{5.100}$$

Components of the vector $G = \{g_1, g_2, g_3\}$ in x, y, z coordinates can be written

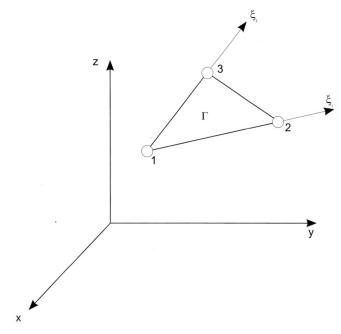

Fig. 5.18. 3-D triangular boundary element

$$g_1 = \frac{\partial y}{\partial \xi_1}\frac{\partial z}{\partial \xi_2} - \frac{\partial y}{\partial \xi_2}\frac{\partial z}{\partial \xi_1} \quad g_2 = \frac{\partial z}{\partial \xi_1}\frac{\partial x}{\partial \xi_2} - \frac{\partial z}{\partial \xi_2}\frac{\partial x}{\partial \xi_1}$$
$$g_3 = \frac{\partial x}{\partial \xi_1}\frac{\partial y}{\partial \xi_2} - \frac{\partial x}{\partial \xi_2}\frac{\partial y}{\partial \xi_1}$$
(5.101)

and then

$$|\vec{G}| = \sqrt{g_1^2 + g_2^2 + g_3^2}.$$
(5.102)

To determine the terms of matrices **H** and **G**, we use these formulas

$$H_{kj}^e = \int_{\Gamma_j} N_k \frac{\partial w_{ij}}{\partial v} d\Gamma_j = \int_0^1\int_0^1 N_k \frac{\partial w_{ij}}{\partial v} |\vec{G}| d\xi_1 d\xi_2$$

$$G_{kj}^e = \int_{\Gamma_j} N_k w_{ij} d\Gamma_j = \int_0^1\int_0^1 N_k w_{ij} |\vec{G}| d\xi_1 d\xi_2.$$
(5.103)

The integrals in relations (5.103) are solved by numerical methods, though if the node i lies inside an element (e.g. is in the centre of an element), the distance r_i is zero and the value w_{ij} is infinite. This point is singular and usual numerical methods are not able to determine the terms of the matrix **G** (all terms of the matrix **H** equal zero). To overcome this problem, a method analogous to that designed by Telles and Brebbia (1980) for plasticity calculations can be used.

5.1.4.6
Unsteady Flow Solution

The planar unsteady flow with a confined surface is governed by this differential equation

$$T_x \frac{\partial^2 \Phi}{\partial x^2} + T_y \frac{\partial^2 \Phi}{\partial y^2} = S \frac{\partial \Phi}{\partial t} \tag{5.104}$$

(for details see Chap. 2). The solution by the boundary element method is again found by a transformation using Eq. (5.86) and the result is

$$\frac{\partial^2 \Phi}{\partial \tilde{x}^2} + \frac{\partial^2 \Phi}{\partial \tilde{y}^2} = \frac{S}{T_x} \frac{\partial \Phi}{\partial t}. \tag{5.105}$$

Two variants are used for a further solution by the boundary element method. The first variant, which can be called traditional, is based on the same principle as the finite element method. That means we use FDM to solve the time dependence. The time derivative on the right side of Eq. (5.105) is replaced by a difference and the entire equation becomes

$$\frac{\partial^2 \Phi}{\partial \tilde{x}^2} + \frac{\partial^2 \Phi}{\partial \tilde{y}^2} - \frac{S}{T_x \Delta t} u_1 = -\frac{S}{T_x \Delta t} u_0. \tag{5.106}$$

Thereafter, we use weight functions that are the fundamental solution of the equation below

$$\frac{\partial^2 u}{\partial \tilde{x}^2} + \frac{\partial^2 u}{\partial \tilde{y}^2} - \frac{S}{T_x \Delta t} u = 0. \tag{5.107}$$

The fundamental solution has this form

$$w = \frac{1}{2\pi} K_0 \left(r \sqrt{\frac{S}{T_x \Delta t}} \right), \tag{5.108}$$

where K_0 is the modified Bessel function of the second kind. After using the inverse formulation, the basic equation has a new form

$$c_k u_k + \int_\Gamma u \frac{\partial w_k}{\partial v} d\Gamma + \int_\Omega \frac{S}{T_x \Delta t} u_0 w_k d\Omega = \int_\Gamma q w_k d\Gamma. \tag{5.109}$$

A matrix variant of this equation is

$$\mathbf{H u} + \mathbf{C u_0} = \mathbf{G q}, \tag{5.110}$$

where the terms of matrices **H**, **G** and **C** are given by these formulas

$$H_{kj} = \int_{\Gamma_j} \frac{\partial w_{kj}}{\partial v} d\Gamma_j + c_k \delta_{kj} \quad G_{kj} = \int_{\Gamma_j} w_{kj} d\Gamma_j \quad C_{kj} = \int_{\Omega} \frac{S}{T_x \Delta t} w_{kj} d\Gamma_j. \tag{5.111}$$

It has to be stressed that this variant uses an integration over a domain in every time step to set the terms of the matrix **C**. Hence, it is again needed to split the domain into subdomains as we did in the case of a planar flow.

The second variant is based on a choice of a new weight function that depends on time and on time integration of the governing Eq. (5.105) (see Brebbia and Walker 1980; Kovarik et al. 1987). After application thereof, together with the inverse formulation in the governing equation, we obtain

$$\int_0^t \int_\Omega \left(\Delta w_k + \frac{S}{T_x} \frac{\partial w_k}{\partial t} \right) u d\Omega dt - \left[\int_\Omega u w_k d\Omega \right]_{t=0} = \int_0^t \int_{\Gamma_2} u \frac{\partial w_k}{\partial v} d\Gamma_2 dt$$

$$+ \int_0^t \int_{\Gamma_1} \bar{u} \frac{\partial w_k}{\partial v} d\Gamma_1 dt - \int_0^t \int_{\Gamma_1} q w_k d\Gamma_1 dt - \int_0^t \int_{\Gamma_2} \bar{q} w_k d\Gamma_2 dt. \tag{5.112}$$

The fundamental solution is a solution of the following equation

$$\Delta w_k + \frac{S}{T_x} \frac{\partial w_k}{\partial t} = \delta(x - x_i) \sqrt{\Lambda} \delta(y - y_i) \tag{5.113}$$

and it has this form

$$w_k = \frac{1}{4\pi T_x t} e^{-\frac{Sr^2}{4\pi T_x t}} \quad \frac{\partial w_k}{\partial v} = \frac{DS}{8\pi T_x^2 t^2} e^{-\frac{Sr^2}{4\pi T_x t}}. \tag{5.114}$$

After substitution of Eq. (5.114) into (5.112) and after the assumption that the rate of change of values u, q in time is lower than that of the value w, a new form of the basic equation originates

$$c_i u_i - \left[\int_\Omega u w_k d\Omega \right]_{t=0} = \int_{\Gamma_2} u \int_0^t \frac{\partial w_k}{\partial v} dt d\Gamma_2 + \int_{\Gamma_1} \bar{u} \int_0^t \frac{\partial w_k}{\partial v} dt d\Gamma_1$$

$$- \int_{\Gamma_1} q \int_0^t w_k dt d\Gamma_1 - \int_{\Gamma_2} \bar{q} \int_0^t w_k dt d\Gamma_2. \tag{5.115}$$

The problem can be simplified by the presumption of homogeneity in the initial condition. This means that $u = 0$ if $t = 0$. This eliminates the last integral over volume on the left side of the equation and it is not necessary to divide the domain. We do not even have to use a numerical integration for integrals over time on the right side of Eq. (5.115) and we analytically obtain

$$\int_0^t w_k dt = \frac{1}{4\pi T_x} \int_a^\infty \frac{e^{-x}}{x} dx = \frac{1}{4\pi T_x} \text{Ei}(a) \quad a = \frac{Sr_{ij}^2}{4T_x t}$$

$$\int_0^t \frac{\partial w_k}{\partial v} dt = -\frac{D}{2\pi T_x r_{ij}^2} e^a, \tag{5.116}$$

where, as in the case of Eq. (5.77), r_{ij} is the distance of the point i from the points on the element j and D is the perpendicular distance between the point i and the element j. Integrals over the elements must be done numerically.

5.1.4.7
Solution Notes

We used approaches from FEM in the boundary element method mainly regarding the choice of base functions. Hee, any resemblance between these two methods ends. In contrast to FEM, the boundary element method has less equations; however, the final matrix is always non-symmetric, is not positively definite and is full. Thus, the solution methods applied in FEM or in the finite differences method cannot be used here. We have to use general solution methods modified for a solution of matrices that are not positively definite (such as the pivoting). The advantage of a lower number of equations is only illusory because it saves neither the time for the solution nor the memory of the computer.

On the other hand, the fact that we divide only the boundary of the domain and the domain into zones speeds up the setting-up of the model. In contrast to elements of FEM, even the simplest elements of the boundary element method work with potential's exterior normal derivative. Flux through elements can be easily determined from these derivatives as well as the balance of inflows and outflows of the whole domain or of a single zone.

5.1.4.8
Inclusion of Boundary Conditions

The inclusion of boundary conditions in this method differs from the two previous methods. The final equation must be rearranged before the solution itself because it has the form of Eq. (5.76). The derivation of the method implies that either the value of a potential u or the value of an exterior normal derivative q must be known. In this case, the number of equations is identical with the number of unknown variables and the problem is solvable. Equation (5.76) is rearranged to

$$\mathbf{A}x = d \quad d = \mathbf{B}y, \tag{5.117}$$

where the matrix $\mathbf{A}$ is composed of columns of the matrices $\mathbf{H}$ and $\mathbf{G}$ that correspond to the unknown potential's values or to the values of the derivative. The matrix $\mathbf{B}$ is composed of columns that correspond to the given boundary conditions of the 1st and 2nd kind.

The inclusion of a boundary condition of the 3rd kind is simple too. When we use the (2.120) form of a boundary condition, the flux after the (5.57) transformation is

$$q = \frac{k_0}{T_x b_0}(\overline{H} - u). \tag{5.118}$$

When relation (5.76) is written in components instead of the matrix form, we have

$$\sum_{j=1}^{N} H_{kj} u_j = \sum_{j=1}^{N} G_{kj} q_j. \tag{5.119}$$

Now q_j is eliminated by a substitution from Eq. (5.118) and the previous relation is derived to

$$\sum_{j=1}^{N} H_{kj} u_j = \sum_{j=1}^{N} G_{kj} \frac{k_0}{T_x b_0} (\overline{H} - u_j). \tag{5.120}$$

We can now clearly see that the column of matrix **A**, which corresponds to the given boundary condition, is

$$A_{kj} = H_{kj} + G_{kj} \frac{k_0}{T_x b_0}. \tag{5.121}$$

On the right side of Eq. (5.117), we have

$$d_k = \frac{k_0 \overline{H}}{T_x b_0} \sum_{j=1}^{N} G_{kj}. \tag{5.122}$$

The inclusion of effects of a point source can be considered as the inclusion of boundary conditions. Even here the manner of doing it differs from the previous methods. Equation (5.56) can be prepared for the inclusion by deriving it using formula (2.108) from Chap. 2

$$T_x \frac{\partial^2 \Phi}{\partial x^2} + T_y \frac{\partial^2 \Phi}{\partial y^2} + \sum_{l=1}^{N_Q} Q_l \delta(x - x_{0l}, y - y_{0l}) = 0, \tag{5.123}$$

where Q_l are yields of the sources and x_0, y_0 are the coordinates of the sources. After transformation of Eq. (5.57), we obtain

$$\frac{\partial^2 \Phi}{\partial \tilde{x}^2} + \frac{\partial^2 \Phi}{\partial \tilde{y}^2} + \sum_{l=1}^{N_Q} \frac{Q_l}{T_x} \delta(\tilde{x} - \tilde{x}_{0l}, \tilde{y} - \tilde{y}_{0l}) = 0. \tag{5.124}$$

Let us now apply the inverse transformation to Eq. (5.124). The result is the basic relation of the boundary element method which we obtained by an approach analogous to that in Sect. 5.1.4.1

$$c_k u_k + \int_\Gamma u \frac{\partial w_k}{\partial \nu} d\Gamma = \int_\Gamma q w_k d\Gamma + \sum_{l=1}^{N_Q} \frac{Q_l}{T_x} w_{kl}, \tag{5.125}$$

where w_{kl} is the weight function between the point k and the source l. By splitting the domain into elements, we obtain this equation

$$c_k u_k + \sum_{j=1}^{N} u_j \tilde{H}_{kj} = \sum_{j=1}^{N} q_j G_{kj} + \sum_{l=1}^{N_Q} \frac{Q_l}{T_x} w_{kl}. \tag{5.126}$$

The last term of Eq. (5.126) represents the influence of point sources. It can be seen that the point sources are not a part of the division of the net of elements. This is a huge difference compared to the previous methods, where the sources are included in the net and any change in the position of the source means a new setting-up of the net. In the boundary element method the position of the source does not influence the net.

The function w_{kl} is given by Eq. (5.66) for a planar case. It is the logarithmic function of the distance between the points k and l. For $k \to l$, obviously $w_{kl} \to \infty$ and we can see that the base functions behave as the Dupuit equation of a steady well recharge (see Mucha and Shestakov 1987). This gives a great advantage to this method because it enables a perfect match of groundwater level in the neighbourhood of a source. Previous methods had to solve this problem by a radical thickening of the net in the neighbourhood of the source or by some empirical corrections.

5.1.5
Dual Reciprocity Method (DRM)

This method has developed from BEM in order to remove the major disadvantages such as the need of a division of a domain into subdomains in case of a planar inflow (see Partridge et al. 1992).

The DR method was first used in a solution of the Poisson equation which has this general form

$$\Delta \Phi = b. \tag{5.127}$$

Its solution can be expressed as the sum of the solution of a homogenous equation and a particular solution $\hat{\Phi}$. As finding this solution is very difficult, the method replaces the solution $\hat{\Phi}$ by a sequence of particular solutions $\hat{\Phi}_i$. The right side of Eq. (5.127) can be approximated by

$$b = \sum_{i=1}^{N+L} \alpha_i f_i, \tag{5.128}$$

where N is the number of boundary nodes and L is the number of inner points. α_i is a set of unknown coefficients and f_i are approximation functions. The particular solutions from the sequence must fulfil these equations

$$\Delta \hat{\Phi}_i = f_i. \tag{5.129}$$

The approximation functions f_i can be chosen as

$$f_i = 1 + r_i + r_i^2 + r_i^3 + \ldots + r_i^m. \tag{5.130}$$

The sequence of particular solutions $\hat{\Phi}_i$ has the form of a series

$$\hat{\Phi}_i = \sum_{k=0}^{m} \frac{r_i^{k+2}}{(k+2)^2}, \tag{5.131}$$

and the exterior normal derivative is

$$\frac{\partial \hat{\Phi}_i}{\partial v} = \frac{\partial r_i}{\partial v} \sum_{k=0}^{m} \frac{r_i^k}{k+2}. \tag{5.132}$$

If we substitute relations (5.128) and (5.129) to Eq. (5.127), we obtain the governing equation

$$\Delta \Phi = \sum_{i=1}^{N+L} \alpha_i (\Delta \hat{\Phi}_i). \tag{5.133}$$

Then we apply approaches similar to BEM and afterwards we acquire

$$c_k u_k + \int_\Gamma \frac{\partial w}{\partial v} u d\Gamma - \int_\Gamma w \frac{\partial u}{\partial v} d\Gamma = \sum_{i=1}^{N+L} \alpha_i \left(c_k \hat{\Phi}_{ki} + \int_\Gamma \frac{\partial w}{\partial v} \hat{\Phi}_i d\Gamma - \int_\Gamma w \frac{\partial \hat{\Phi}_i}{\partial v} d\Gamma \right). \tag{5.134}$$

The unknown coefficients in Eq. (5.128) can be determined from the known values b in the $N + L$ points.

$$\alpha_i = \sum_{j=1}^{N+L} F_{ij}^{-1} b_j, \tag{5.135}$$

where we denoted $\mathbf{F}^{-1}$ the inverse matrix to the matrix of approximation functions f_{ij}. A matrix form of Eq. (5.134) is

$$\mathbf{Hu} - \mathbf{Gq} = (\mathbf{H\hat{u}} - \mathbf{G\hat{q}})\mathbf{F}^{-1}\mathbf{b}. \tag{5.136}$$

The left side of Eq. (5.136) is totally identical with the solution of homogenous Eq. (5.76). On the right side, there are matrices $\mathbf{H}$ and $\mathbf{G}$ along with the matrices of particular solutions. The matrices of particular solutions can be easily set up using relations (5.131) and (5.132). The vector $\mathbf{b}$ consists of given values of the planar inflow in every node of the net and in inner points. The DR method requires a certain number of inner points to be given where the inflow value b is set. The inclusion of boundary conditions is the same as in Sect. 5.1.4.8 because the DR method is a variant of BEM.

5.2
Flow in Unsaturated Zone

The flow is governed by the basic differential equation which was derived in Chap. 2.3. In contrast to the saturated zone, there are always non-linear relations, which makes the solution much more difficult. One has to pick a method so that it converges (sometimes there is no such method). Owing to the complications, everyone's attention was focused mainly on 1-D models of vertical transport of moisture in a soil column. Later, solutions of planar cases emerged but exclusively in a vertical profile. 3-D models are still very rare.

We showed in Chap. 2 that there are two types of governing equation – the diffusion type and the capacity type. The former is the easier to solve because there is no hysteresis, thus it is used in most analytic solutions. It has, however, one main disadvantage compared with the latter: it cannot tackle an interaction of the saturated and the unsaturated zone.

5.2.1
Analytic Solutions

As in the saturated zone, the analytic solutions were the first solutions in the unsaturated zone as well. They were direct analytic solutions of the Richards equation (see Chap. 2) particularly of the diffusion type of that equation for a 1-D case. This equation is strongly non-linear and its analytic solution can be obtained only for a steady flow and in some unsteady cases. Otherwise, approximate analytic solutions and iteration methods are applied. A whole series of these solutions is known and can be found in (Kutilek 1984). Nowadays, the numerical methods are overwhelmingly used.

5.2.2
Finite Differences Method

This was the first purely numerical method applied in modelling of a flow in the unsaturated zone. It is the most common method, since the models in the unsaturated zone are primarily one-dimensional models of infiltration into covering soil layer or planar models of water transport in a vertical soil profile. These imply lower requirements in the match of a complex shape of domains, which is the main disadvantage of this method.

The solution uses the capacity variant of the governing equation (see Chap. 2.3)

$$C\frac{\partial H}{\partial t} = \frac{\partial}{\partial z}\left[K(H)\left(\frac{\partial H}{\partial z}-1\right)\right]. \tag{5.137}$$

We can again use the Taylor series as in Sect. 5.1.2 to approximate derivatives in Eq. (5.13). We obtain different schemes, depending on the approximation of time. The only difference lies in the non-linearity of coefficients K and C. This non-linearity can be dealt with by iterations in every time step as in FEM (see below). A more complex time schema is used instead of iterations (see Haverkamp et al. 1977) and the values of K and C are calculated by various averaging schemes. The simplest of the schemes is an implicit schema with an explicit linearisation having this form

$$C_i^j \frac{H_i^{j+1}-H_i^j}{\Delta t} = \frac{1}{\Delta z}\left[K_{i+1/2}^j\left(\frac{H_{i+1}^{j+1}-H_i^{j+1}}{\Delta z}-1\right)-K_{i-1/2}^j\left(\frac{H_i^{j+1}-H_{i-1}^{j+1}}{\Delta z}-1\right)\right], \tag{5.138}$$

where index i stands for a spatial coordinate and the index j for the time step. Subsequently, the values of parameters can be calculated as an average

$$K_{i+1/2} = \frac{1}{2}(K_i + K_{i+1}) \tag{5.139}$$

or as a harmonic average

$$K_{i+1/2} = \frac{2 \cdot K_i K_{i+1}}{K_i + K_{i+1}}, \tag{5.140}$$

or as a geometric average (Haverkamp and Vauclin 1979)

$$K_{i+1/2} = \sqrt{K_i K_{i+1}}. \tag{5.141}$$

A detailed review of the schemes can be found in Kutilek (1984).

5.2.3
Finite Element Method

The finite element method began to be used for a flow in the unsaturated zone in the 1970s. The Galerkin variant (weight and base functions are identical) is used for deriving the basic equations of this method in a majority of cases. The derivations will be shown on an example of a one-dimensional vertical model, which is the most frequently used model. The simplest element is an element with linear base functions and two nodes. If we use the capacity variant, the unknown parameter is the pressure head H which is in every element approximated by base functions. The approximation is

$$H(z,t) \cong \sum_{i=1}^{n} H_i(t) N_i(z), \tag{5.142}$$

where $H_i(t)$ are unknown values of the pressure head in nodes. $N_i(z)$ are the base functions and n is the number of nodes in an element. If we use the weak solution of a basic differential equation (see Chap. 4.3) in Eq. (5.137), we obtain this relation

$$\int_\Omega \left[K(H) \frac{\partial H}{\partial z} \frac{\partial N_j}{\partial z} + K(H) \frac{\partial N_j}{\partial z} + C(H) \frac{\partial H}{\partial z} N_j \right] d\Omega + \int_{\Gamma_2} q N_j d\Gamma = 0 \tag{5.143}$$

$$j = 1, 2, \ldots, N,$$

where N is the total number of nodes in the net. After substitution of Eq. (5.142) into Eq. (5.143) and after a further derivation, we obtain a system of non-linear differential equations

$$\mathbf{B}\Delta + \mathbf{M} \frac{d\Delta}{dt} + \mathbf{F} = 0, \tag{5.144}$$

where

$$B_{ij} = \int_\Omega K(H) \frac{\partial N_i}{\partial z} \frac{\partial N_j}{\partial z} d\Omega$$

$$M_{ij} = \int_\Omega C(H) N_i N_j d\Omega \tag{5.145}$$

$$F_i = \int_\Omega K(H) \frac{\partial N_i}{\partial z} d\Omega + \int_{\Gamma_2} N_i q d\Gamma.$$

The vector Δ contains the values of the pressure head H_i in every node. We solve the time dependence by the finite differences method by introducing

$$\Delta = (1-\xi)\Delta_1 + \xi\Delta_2, \tag{5.146}$$

thereafter we gain a recurrent formula

$$\left(\xi \mathbf{B} + \frac{1}{t_2 - t_1} \mathbf{M}\right) \Delta_2 = \left[(\xi - 1)\mathbf{B} + \frac{1}{t_2 - t_1} \mathbf{M}\right] \Delta_1 + (1-\xi)\mathbf{F}(t_1) + \xi \mathbf{F}(t_2). \tag{5.147}$$

This formula represents a system of non-linear equations that has to be solved in each time step. A different choice of parameter ξ results in various schemes (e.g. for $\xi = 0.5$ it is the Crank-Nicholson schema). The system is solved by an iteration method derived from the Newton-Raphson method (see Segol 1976; Khaleel and Yeh 1985). The coefficients K and C are determined from relations in Chap. 2.2 in every node with values H being from the middle of a time step. The coefficients are approximated by the base functions

$$K(H,z) \cong \sum_{i=1}^{n} K_i(H) N_i(z) \quad C(H,z) \cong \sum_{i=1}^{n} C_i(H) N_i(z). \tag{5.148}$$

An initial estimate of the pressure head's value for the iteration in the $(k+1)$th time step is

$$\Delta_0^{k+1} = \Delta^k + \frac{\tau_k}{2\tau_{k-1}}(\Delta^k - \Delta^{k-1}) \quad \tau^k = t_{k+1} - t_k. \tag{5.149}$$

The mth iteration step is

$$\Delta_m^{k+1} = \frac{1}{2}(\Delta_{m-1}^{k+1} + \Delta^k). \tag{5.150}$$

Using these estimated values in a system of linear equations, Eq. (5.147) yields another estimate of Δ^{k+1}.

The simplest element is a linear one with two nodes and two base functions in the form of a linear polynomial

$$N_1 = \frac{1}{L}(z_2 - z) \quad N_2 = \frac{1}{L}(z - z_1) \tag{5.151}$$

where L is the element's length and z_1, z_2 are the z-coordinates of the nodes, respectively. The setup of the element's matrices using formula (5.145) is again very easy because it is integration of a constant function. Now we have new relations for the coefficients K and C

$$K(H,z) = \frac{1}{L}(K_1 z_2 - K_2 z_1) + \frac{z}{L}(K_2 - K_1)$$
$$C(H,z) = \frac{1}{L}(C_1 z_2 - C_2 z_1) + \frac{z}{L}(C_2 - C_1). \quad (5.152)$$

The basic matrices are of a (2 × 2) type and have the following terms

$$B_{11} = B_{22} = \frac{1}{2L}(K_1 + K_2) \quad B_{12} = B_{21} = -B_{11}$$

$$M_{11} = \frac{L}{12}(2C_1 + C_2) \quad M_{22} = \frac{L}{12}(2C_2 + C_1)$$

$$M_{12} = M_{21} = \frac{L}{12}(C_1 + C_2) \quad (5.153)$$

$$F_1 = \frac{1}{2}(K_1 + K_2) \quad F_2 = -\frac{1}{2}(K_1 + K_2).$$

The final system of equations is tridiagonal and can be easily and quickly solved. A very interesting type of element with a polynomial of a higher order was proposed by van Genuchten (1983). He used an element with the Hermite interpolation polynomial of the 3rd order which is from the $W_2^{(2)}$ space. It gives continuous values of the 1st derivatives. Due to this property, we obtain a more precise calculation of the velocity of the flow, which is of an immense importance to models of transport of pollutants. It is advantageous to transform an element into a unit element with the help of the formula below, as integrals in Eq. (5.145) must be calculated by numerical methods

$$\xi = -1 + \frac{2}{L}(z - z_1) \quad (5.154)$$

where L is again the length of the element. The base functions have a form of the Hermite polynomial

$$N_{01} = \frac{1}{4}(\xi - 1)^2(\xi + 2) \quad N_{02} = \frac{1}{4}(\xi + 1)^2(\xi - 2)$$
$$N_{11} = \frac{1}{4}(\xi - 1)^2(\xi + 1) \quad N_{12} = \frac{1}{4}(\xi + 1)^2(\xi - 1) \quad (5.155)$$

and the pressure head is approximated as follows

$$H(z,t) = \sum_{i=1}^{2} H_i N_{0i} + \frac{\partial H_i}{\partial z} N_{1i}. \quad (5.156)$$

We use Eq. (5.152) to interpolate the values K, C and afterwards we obtain these basic relations

$$B_{11} = \frac{1}{2}\int_{-1}^{1}[K_1 + K_2 + (K_2 - K_1)\xi]\frac{\partial N_{01}}{\partial z}\frac{\partial N_{01}}{\partial z}|J|d\xi$$

$$B_{12} = \frac{1}{2}\int_{-1}^{1}[K_1 + K_2 + (K_2 - K_1)\xi]\frac{\partial N_{01}}{\partial z}\frac{\partial N_{11}}{\partial z}|J|d\xi$$

$$C_{ij} = \frac{1}{2}\int[C_1 + C_2 + (C_2 - C_1)\xi]N_{0i}N_{0j}|J|d\xi$$

$$F_i = \frac{1}{2}\int_{-1}^{1}[K_1 + K_2 + (K_2 - K_1)\xi]\frac{\partial N_{0i}}{\partial z}|J|d\xi.$$

(5.157)

The vector Δ contains values of the pressure head and of the pressure head's derivative in nodes. Equation (5.158) represents a value $|J|$ of the Jacobian of transformation [Eq. (5.154)]. For this element

$$|J| = \frac{\partial z}{\partial \xi} = \frac{L}{2}. \tag{5.158}$$

Using the element, the number of unknown parameters doubles and the order of the matrices **B** and **M** increases along with the bandwidth. Additionally, we have to do numerical integration to set up the system of equations. This fact causes the solution to be much slower than by the linear element. A new method offered an interesting way in which to overcome the disadvantages and still keep the derivatives of the 1st order continuous. It is the collocation method (see below) which is used instead of the Galerkin variant to derive the basic equation.

A planar case is mostly a case of a water flow in the unsaturated zone in a vertical profile. Here, the most suitable elements are the isoparametric elements as described in Sect. 5.1.3. Mostly, the four- and eight-node elements are applied. The elements with eight nodes are more fitting because the shape of the domain usually has one dominant dimension (mainly the length of a profile).

5.2.4
Collocation Method

The collocation method is successfully used in a water flow in the unsaturated zone as an alternative to the Galerkin method (see Allen and Murphy 1985; Kovarik 1991). It has a special set of base functions which is similar to the one in the previous paragraph. This method is not very usual and therefore we summarise its differences compared to the commonly used methods. Above all, it is formulated with the help of a pressure head's increment and not with its absolute value. It requires the base functions to be polynomials of a higher order. This positively influences the solution's precision. Furthermore, it does not need any integration when setting up the local matrix of an element, which accelerates the solution even more. On the other hand, the final matrix of the system is non-symmetric, which increases the number of operations needed for its solution. This disadvantage is not so distinct in a 1-D case because the matrices are tridiagonal and can be solved easily. This method has a few problems with the net in 2-D as it does not enable such variability as FEM. Considering all the pros and cons, we can recommend this method in a one-dimensional case, but one must expect problems in more-dimensional cases.

References

Allen MB, Murphy C (1985) A finite-element collocation method for variably saturated flows in porous media. Num. Meth. Part. Diff. Eq. 3:229–239

Banský V, Kovarik K (1978) Riešenie trojrozmerného nestacionárneho prúdenia podzemnej vody metódou konecných prvkov. Vodohosp. cas. 26:293–314

Bear J (1972) Dynamics of fluids in porous media. American Elsevier, New York

Brebbia CA, Walker J (1980) Boundary element techniques in engineering. Newness-Butterworth, London

Brebbia CA, Telles JFC, Wrobel LC (1984) Boundary element techniques, Springer, Berlin Heidelberg New York

Broz P, Procházka P (1987) Metoda okrajových prvku v inzenýrské praxi. SNTL, Prague

Connor JJ, Brebbia CA (1976) Finite element techniques in fluid flow. Butterworth, London

Hálek V, Švec J (1979) Groundwater hydraulics. Academia, Prague

Haverkamp R, Vauclin M, Touma J, Wierenga PJ, Vachaud G (1977) A comparison of numerical simulation models for one-dimensional infiltration. Soil Sci. Am. J. 41:285–293

Haverkamp R, Vauclin M (1979) Note on estimating finite difference interblock hydraulic conductivity values for transient unsaturated flow problems. Water Res. Res. 15:180–187

Irons BM (1970) A frontal solution program for finite element analysis. Int. J. for Num. Meth. in Engng. 2:52–61

Jaswon MA (1963) Integral equation methods in potential theory I. Proc. Roy. Soc. A. 275:23–32

Kazda I (1983) Proudení podzemní vody, rešení metodou konecných prvku. SNTL, Prague

Kazda I (1989) Prostorové konecné prvky a jejich vhodnost pro resení trojrozmerného proudení podzemní vody, Acta Polytechnica CVUT Prague, 10(I,3)

Khaleel R, Yeh TC (1985) Galerkin-finite element program for simulating unsaturated flow in porous media, Ground Water. 23:90–96

Kolar V, Kratochvil J, Leitner F, Zenisek A (1972) Výpocet plošných a prostorových konstrukcí metodou konecných prvku. SNTL, Prague

Kovarik K (1991) Prenos ionov roznych latok cez nenasytenu zonu, PhD thesis, SAV, Bratislava

Kovarik K (1993) Numerické metódy pre riešenie rovníc prúdenia podzemnej vody a prenosu látok v pórovom prostredí, habil. thesis CVUT, Prague

Kovarik K, Drahos M, Sinkova M (1987) Moderné metódy matematického modelovania v hydrogeológii. Research report R-52-547-173, archive IGHP, Zilina

Kovarik K (1978) Rešení hydrodynamické disperze metodou konecných prvku. Vodohosp. cas. 26:204–216

Kutílek M (1984) Vlhkost pórovitých materiálu. SNTL, Prague

Luckner L, Šestakov VM (1976) Modelirovanie geofiltracii. Nedra, Moscow

Martin A, Rodriguez I, Alarcon E (1980) Mixed elements in the boundary theory. In: Brebbia CA (ed) New developments in boundary element methods. CML Publications, Southampton, pp. 34–42

Mucha I, Šestakov VM (1987) Hydraulika podzemných vôd. Alfa, Bratislava

Partridge PW, Brebbia CA, Wrobel LC (1992) The dual reciprocity boundary element method. CM Publications, Southampton

Peaceman DW, Rachford Jr HH (1955) The numerical solution of parabolic and elliptic differential equations. J. Soc. Indust. App. Math. 3

Remson, Hornberger, Moltz (1970) Numerical methods in subsurface hydrology. J. Wiley & Sons, New York

Segol G (1976) A three-dimensional galerkin finite element model for the analysis of contaminant transport in saturated-unsaturated porous media. In: Finite elements in water resources. Pentech Press, London

Telles JCF, Brebbia CA (1980) The boundary element method in plasticity. In: Brebbia CA (ed) New developments in boundary element methods. pp. 295–317 CML Publications, Southampton

van Genuchten MT (1983) An hermitian finite element solution of the two-dimensional saturated-unsaturated flow equation. Adv. Water Res. 6:106–111

6
Mathematical Models of Transport of Miscible Pollutants

A transport of miscible pollutants is governed by a differential equation of dispersion (see Chap. 3). In contrast to the groundwater flow models, the models of pollutant transport have to deal with the effects of an advective transport. This makes the solution difficult and, as we will see later, a new error emerges, which causes an instability of the front of a transition zone by high velocities (that is when the advective term has a great influence). To assess the influence, the Peclet number is introduced

$$Pe = \frac{vL}{D}, \qquad (6.1)$$

where v is the velocity of a flow, L is a characteristic length; in porous media we use an average diameter of grains (d_{50} value from the gradation curve); in modelling the length of a cell of the net is used instead and this Peclet number is called local. In the original definition D is the diffusion coefficient, but currently the dispersion coefficient is being used (for more details see Chap. 3).

Obviously, the higher the velocity, the larger is the Peclet number. When the velocity equals zero, only a molecular diffusion comes into play and $Pe = 0$. The other limit case is a pure advective transport when $D = 0$ and $Pe \to \infty$.

As mentioned above, classical methods based on the weighted residuals method often fail when applied to pollutant transport. To solve transport problems, the finite element method in the Galerkin variant is the most frequently used of all classical methods. The advective term causes an instability of the pollution front and also a so-called numerical dispersion. This numerical dispersion is a phenomenon describing an artificial spread of pollutants that has its origin in the numerical method used. It means that the use of a numerical method results in a larger spread than there actually is. By pollution front instability we mean oscillations close to the front. Owing to this instability, even negative concentrations can come out of the solution. Both these phenomena can be illustrated on a one-dimensional example (see Fig. 6.1). These phenomena act against each other, which means that when we want to decrease the numerical dispersion, we will automatically increase the oscillations, and vice versa. A major

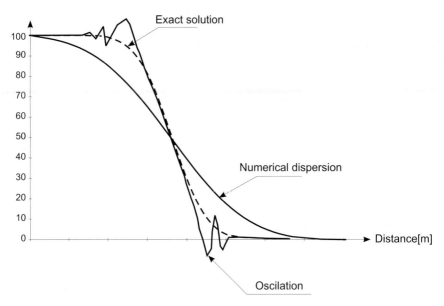

Fig. 6.1. Numerical dispersion and instability of solution

advection is present when $Pe > 100$, and different approaches have to be used to correct these problems (see Huyakorn and Nilkuha 1979). In the weighted residuals method the weight functions are corrected. As the weight and base functions are no longer identical, this variant is termed the Petrov-Galerkin method.

These problems led to the development of a whole series of methods that use different approaches. We can call them all particle methods. They try to overcome the problems with the transport by emitting particles. The main two are the method of characteristic curves and the random walk method.

6.1
Basic Methods

In this section we concentrate on a description of methods that are nowadays used in pollutant transport problems. We do it by the simplest example of a governing equation without considering any influences such as the sorption of pollutants on the soil complex. In the next sections we will show possibilities of how to cope with these problems.

The finite differences method is missing from the list because this method is now no longer used on its own. It is often used combined with particle methods, as are the two mentioned above.

6.1.1
Analytic Solution

The first solutions that appeared were analytic, similarly to Chap. 5. These solutions are restricted to cases with simple boundary conditions and simple geometric shapes of a domain. Additionally, another restriction occurred, since the direct solutions assumed that the transport takes place in a homogenous field of velocities (the velocities do not change in the entire domain). This is the greatest obstacle that hinders the use of the solutions and they are used only to test numerical models.

The most popular analytic solutions come from the works of Ogata and Banks (1961) and Bear (1972). These are the solutions of a transport in a semiinfinite horizontal strip. The solution stems from a basic differential equation of a one-dimensional transport (for details see Chap. 3)

$$\frac{\partial C}{\partial t} = D\frac{\partial^2 C}{\partial x^2} - v_p\frac{\partial C}{\partial x}, \tag{6.2}$$

where D is the dispersion coefficient and v_p is the actual (porous) velocity of a flow parallel with the x-axis and it is constant in the entire domain. The solution of this equation with a boundary condition $C = 0$ at time $t = 0$ and $C = C_0$ for $x = 0$ (the Dirichlet boundary condition of the first kind) is

$$\frac{C(x,t)}{C_0} = \frac{1}{2}\left[\text{erfc}\left(\frac{x-v_p t}{2\sqrt{Dt}}\right) + e^{\left(\frac{v_p x}{D}\right)}\text{erfc}\left(\frac{x+v_p t}{2\sqrt{Dt}}\right)\right]. \tag{6.3}$$

The function erfc(x) is complementary to the distribution function of a normal distribution erf(x)

$$\text{erfc}(x) = 1 - \text{erfc}(x). \tag{6.4}$$

The distribution function is

$$\text{erf}(x) = \frac{1}{\sqrt{2\pi}}\int_0^x e^{-\frac{u^2}{2}} du. \tag{6.5}$$

In calculations, a power series expansion of the distribution function is used (Abramovitz and Stegun 1972)

$$\text{erf}(x) = \frac{2}{\sqrt{\pi}}\sum_{n=0}^{\infty}\frac{(-1)^n x^{2n+1}}{n!(2n+1)}. \tag{6.6}$$

6.1.2
Finite Element Method

The foundation of the finite element method is the Galerkin variant of the weighted residuals method like in the groundwater flow in Chap. 5. The weak solution of the basic governing differential equation of pollutant transport looks like this:

$$\int_\Omega \left(D_{km} \frac{\partial N_i}{\partial x_k} \frac{\partial N_j}{\partial x_m} - v_k \frac{\partial N_i}{\partial x_k} N_j \right) C_i d\Omega - \int_\Gamma v_m D_{km} \frac{\partial N_i}{\partial x_m} N_j C_i d\Gamma \\ - \int_\Omega N_i N_j \frac{dC_i}{dt} d\Omega = 0. \tag{6.7}$$

We obtain a set of differential equations in a manner analogous to Chap. 5.1.3

$$\mathbf{HC} + \mathbf{M} \frac{d\mathbf{C}}{dt} + \mathbf{R} = \mathbf{0}, \tag{6.8}$$

where (see Kovarik 1978)

$$H_{ij} = \int_\Omega \left(D_{km} \frac{\partial N_i}{\partial x_k} \frac{\partial N_j}{\partial x_m} - v_k \frac{\partial N_i}{\partial x_k} N_j \right) d\Omega$$

$$R_i = -\int_\Gamma v_m D_{km} \frac{\partial N_i}{\partial x_m} N_j d\Gamma \tag{6.9}$$

$$M_{ij} = -\int_\Omega N_i N_j d\Omega.$$

The system can be solved with the help of a time difference schema (see Chap. 5.1.3)

$$\mathbf{C} = (1-\xi)\mathbf{C}_1 + \xi \mathbf{C}_2 \quad \xi \in \langle 0,1 \rangle \\ \mathbf{C}_1 = \mathbf{C}(t_1) \quad \mathbf{C}_2 = \mathbf{C}(t_2) \quad \Delta t = t_2 - t_1 \tag{6.10}$$

and we obtain a recurrent formula

$$\left(\xi \mathbf{H} + \frac{1}{\Delta t} \mathbf{M} \right) \mathbf{C}_2 = \left[(\xi-1)\mathbf{H} + \frac{1}{\Delta t} \mathbf{M} \right] \mathbf{C}_1 + (1-\xi)\mathbf{R}(t_1) + \xi \mathbf{R}(t_2). \tag{6.11}$$

This relation, together with the known initial condition, $C = C_0$ at time $t = t_0$, yields the concentration values in mesh points in every time step. By a different choice of ξ-parameter, we can obtain various schemes. If $\xi = 1$, we have the implicit schema which is the most stable and the most popular one. Compared to the groundwater flow models, the matrix $\mathbf{H}$ is not symmetrical, thanks to the advective term. It causes difficulties which require the use of specialised programs for solving large non-symmetrical sets of equations, unlike the Choleski method which is suitable only for symmetrical sets.

The relations above are the Galerkin method used together with the finite element method. As mentioned in the first paragraph, this method can fail when the advective term is dominant. Actually, it can be used when the disperse transport is dominant (that is when $Pe < 10$). On other occasions the Petrov-Galerkin method is applied.

This is a modified Galerkin method where the weight and base functions are not identical (see Chap. 4). Eq. (6.7) has this form

$$\int_\Omega \left(D_{km} \frac{\partial N_i}{\partial x_k} \frac{\partial W_j}{\partial x_m} - v_k \frac{\partial N_i}{\partial x_k} W_j \right) C_i d\Omega - \int_\Gamma v_m D_{km} \frac{\partial N_i}{\partial x_m} W_j C_i d\Gamma$$
$$- \int_\Omega N_i W_j \frac{dC_i}{dt} d\Omega = 0, \qquad (6.12)$$

where the weight functions W_j appear (and are different from the base functions N_i). All matrices change in the same manner

$$H_{ij} = \int_\Omega \left(D_{km} \frac{\partial N_i}{\partial x_k} \frac{\partial W_j}{\partial x_m} - v_k \frac{\partial N_i}{\partial x_k} W_j \right) d\Omega$$

$$R_i = -\int_\Gamma v_m D_{km} \frac{\partial N_i}{\partial x_m} W_j d\Gamma \qquad (6.13)$$

$$M_{ij} = -\int_\Omega N_i W_j d\Omega.$$

The actual choice of the weight functions is only in a stage of development. There have been some results published in works of (Huyakorn and Nilkuha 1979) who introduced upstream factors and proposed this form of the weight functions

$$W_1 = \frac{1}{16}[(1+\xi)(3\alpha_1\xi - 3\alpha_1 - 2)][(1+\eta)(-3\beta_2\eta - 3\beta_2 - 2)]$$
$$W_2 = \frac{1}{16}[(1+\xi)(-3\alpha_1\xi + 3\alpha_1 + 2)][(1+\eta)(3\beta_1\eta - 3\beta_1 - 2)] \qquad (6.14)$$
$$W_3 = \frac{1}{16}[(1+\xi)(-3\alpha_2\xi + 3\alpha_2 + 2)][(1+\eta)(-3\beta_1\eta + 3\beta_1 + 2)]$$
$$W_4 = \frac{1}{16}[(1+\xi)(3\alpha_2\xi - 3\alpha_2 - 2)][(1+\eta)(-3\beta_2\eta + 3\beta_2 + 2)].$$

These are the weight functions of an isoparametric element of a general quadrilateral shape. In formulas (6.14), we introduced the upstream factors α_1, α_2, β_1, β_2 for 1-2, 3-4, 2-3, 4-1 sides of an element. The coordinates ξ, η are a unit local system of coordinates in each element, $\xi \in \langle -1,1 \rangle$, $\eta \in \langle -1,1 \rangle$.

6.1.3
Dual Reciprocity Method

We have already discussed this method in Chap. 5 as a variant of the boundary element method which is suitable to solve the Poisson equation. After some derivations it is even fit for a solution of pollutant transport.

Before all else, we have to derive the basic equation of transport with the help of this coordinates' transformation

$$\tilde{x} = x \quad \tilde{y} = y\sqrt{\Lambda} \quad \Lambda = \frac{D_x}{D_y} \tag{6.15}$$

to the form of the Poisson equation

$$\frac{\partial^2 C}{\partial \tilde{x}^2} + \frac{\partial^2 C}{\partial \tilde{y}^2} = \frac{v_x}{D_x}\frac{\partial C}{\partial \tilde{x}} + \frac{v_y}{D_x}\frac{\partial C}{\partial \tilde{y}} + \frac{1}{D_x}\frac{\partial C}{\partial t}. \tag{6.16}$$

The right side of Eq. (6.16) is a sum of three terms b_1, b_2 and b_3 where

$$b_1 = \frac{v_x}{D_x}\frac{\partial C}{\partial \tilde{x}} \quad b_2 = \frac{v_y}{D_x}\frac{\partial C}{\partial \tilde{y}} \quad b_3 = \frac{1}{D_x}\frac{\partial C}{\partial t}. \tag{6.17}$$

We can apply the same approach as in Chap. 5 to the first two terms b_1, b_2. Let us approximate them by functions f_i

$$b_1 = \sum_{i=1}^{N+L} \alpha_{1i} f_i \quad b_2 = \sum_{i=1}^{N+L} \alpha_{2i} f_i. \tag{6.18}$$

N is here the number of nodes in the boundary elements and L is the number of inner points. To set the unknown coefficients α_{1i} and α_{2i} in Eq. (6.18), we need to know the values b_1, b_2 in $N + L$ points. Then the coefficients gain the form

$$\alpha_{1i} = \sum_{j=1}^{N+L} F_{ij}^{-1} b_{1j} \quad \alpha_{2i} = \sum_{j=1}^{N+L} F_{ij}^{-1} b_{2j}. \tag{6.19}$$

After substituting for b_1, b_2 from Eq. (6.17), we obtain

$$\alpha_{1i} = \frac{v_x}{D_x}\sum_{j=1}^{N+L} F_{ij}^{-1} \frac{\partial C_j}{\partial \tilde{x}} \quad \alpha_{2i} = \frac{v_y}{D_x}\sum_{j=1}^{N+L} F_{ij}^{-1} \frac{\partial C_j}{\partial \tilde{y}}. \tag{6.20}$$

Just to be illustrative, let us focus on an equation that has only b_1 on the right side

$$\frac{\partial^2 C}{\partial \tilde{x}^2} + \frac{\partial^2 C}{\partial \tilde{y}^2} = b_1. \tag{6.21}$$

If we use the notation of the boundary element method, we have [see also Eq. (5.136)]

$$\mathbf{HC} - \mathbf{GQ} = \frac{v_x}{D_x}(\mathbf{H\hat{C}} - \mathbf{G\hat{Q}})\mathbf{F}^{-1}\frac{\partial \mathbf{C}}{\partial \tilde{x}}. \tag{6.22}$$

Now we need to express the concentration's derivative in $N + L$ points. To do this we use this formula (see Patridge et al. 1992)

$$\frac{\partial \mathbf{C}}{\partial \tilde{x}} = \frac{\partial \mathbf{F}}{\partial \tilde{x}} \mathbf{F}^{-1} \mathbf{C}. \tag{6.23}$$

After substituting Eq. (6.23) into Eq. (6.22), we obtain

$$\mathbf{HC} - \mathbf{GQ} = \frac{v_x}{D_x}(\mathbf{H\hat{C}} - \mathbf{G\hat{Q}})\mathbf{F}^{-1}\frac{\partial \mathbf{F}}{\partial \tilde{x}}\mathbf{F}^{-1}\mathbf{C}. \tag{6.24}$$

Let us denote

$$\mathbf{A} = \frac{v_x}{D_x}(\mathbf{H\hat{C}} - \mathbf{G\hat{Q}})\mathbf{F}^{-1}\frac{\partial \mathbf{F}}{\partial \tilde{x}}\mathbf{F}^{-1} \tag{6.25}$$

and we have

$$(\mathbf{H} - \mathbf{A})\mathbf{C} - \mathbf{GQ} = 0. \tag{6.26}$$

If we do the same for the b_2 term, the matrix has a new form

$$\mathbf{A} = (\mathbf{H\hat{C}} - \mathbf{G\hat{Q}})\mathbf{F}^{-1}\left(\frac{v_x}{D_x}\frac{\partial \mathbf{F}}{\partial \tilde{x}} + \frac{v_y}{D_x}\frac{\partial \mathbf{F}}{\partial \tilde{y}}\right)\mathbf{F}^{-1}. \tag{6.27}$$

Equation (6.26) stays formally unchanged. The term b_3 includes a derivative of the concentration with respect to time and we can use the same approach as above. We can define

$$\alpha_{3i} = \frac{1}{D_x}\sum_{j=1}^{N+L} F_{ij}^{-1}\frac{\partial C_j}{\partial t}. \tag{6.28}$$

Equation (6.26) changes to

$$(\mathbf{H} - \mathbf{A})\mathbf{C} - \mathbf{GQ} = \mathbf{B}\frac{\partial \mathbf{C}}{\partial t}. \tag{6.29}$$

The matrix $\mathbf{B}$ has this form

$$\mathbf{B} = \frac{1}{D_x}(\mathbf{H\hat{C}} - \mathbf{G\hat{Q}})\mathbf{F}^{-1}. \tag{6.30}$$

Equation (6.30) can be solved by a standard approach (integration over time). Let us introduce new symbols

$$\mathbf{C} = (1 - \xi_c)\mathbf{C}_1 + \xi_c\mathbf{C}_2 \quad \mathbf{Q} = (1 - \xi_Q)\mathbf{Q}_1 + \xi_Q\mathbf{Q}_2, \tag{6.31}$$

where ξ_c, ξ_q are parameters that determine the course of values $\mathbf{C}, \mathbf{Q}$ between times 1 and 2. Let us substitute this into Eq. (6.29) and subsequently we obtain a recurrent formula

$$\begin{aligned}\left(\frac{1}{\Delta t}\mathbf{B} + \xi_c(\mathbf{H} - \mathbf{A})\right)\mathbf{C}_2 - \xi_Q\mathbf{GQ}_2 \\ = \left(\frac{1}{\Delta t}\mathbf{B} - (1 - \xi_c)(\mathbf{H} - \mathbf{A})\right)\mathbf{C}_1 + (1 - \xi_Q)\mathbf{GQ}_1,\end{aligned} \tag{6.32}$$

which can be solved in every time step. In contrast to Eq. (6.11), which solves a similar problem in the finite element method, we approximate also the course of variable Q. This stands for an exterior normal derivative of concentration. The use of the dual reciprocity method in groundwater hydraulics is still in its beginnings. Results that have been acquired by this method so far show an immense stability in time, which allows us to use larger time steps as compared to the finite element method.

6.1.4
Method of Characteristic Curves

This method tries to solve the disperse and the advective part of transport separately. We will discuss a transport without sorption and decay processes not to make things complicated. When the advective term is removed from the basic Eq. (3.13) (see Chap. 3), we have a new basic equation of the dispersion transport

$$\frac{\partial C}{\partial t} = \frac{\partial}{\partial x_i}\left(D_{ij} \frac{\partial C}{\partial x_j}\right). \qquad (6.33)$$

This equation can be solved by the finite differences method (see Reddell and Sunada 1970) as well as by the finite element method. After removal of the advective term, the operator changes to a symmetrical one and the solution itself is almost easy (it is similar to the solution of an unsteady groundwater flow, see Chap. 5.1).

The advective transport solution is carried out by integration over a small domain moving in the direction of the flow (along a characteristic curve). This is done by placing particles uniformly in each element (particles act as a some kind of centres of mass that have a defined concentration of the pollutant). These particles move in the direction of flow. The concentration of the pollutant changes continuously and the change is caused by the disperse solution. The number of particles in the neighbourhood of a mesh point act the other way round, changing the final concentration.

We must have a given initial value of concentration when solving a problem that depends on time. We prepare the usual net and generate a uniform distribution of particles in each element (or block, depending on the method we use for the solution of disperse transport). Every particle has a given certain initial value of concentration according to the initial condition. Thereafter, we choose the time step used in the solution. The solution in each time step can be divided into three stages.

1st stage. We move the particles along the streamline. This means, we change their position using the relations

$$\begin{aligned} x_c(t+\Delta t) &= x_c(t) + \Delta t v_x(x_c, y_c) \\ y_c(t+\Delta t) &= y_c(t) + \Delta t v_y(x_c, y_c). \end{aligned} \qquad (6.34)$$

After the change we determine a new concentration $C_i(t + \Delta t)$ in every mesh point (or block) as an average concentration of all particles in the mesh point's neighbourhood (or in the block).

2nd stage. At this stage the solution of the dispersion part is carried out. This can be different, depending on the method we use. In the solution we use a concentration that is an average of the concentration in the preceding time step and of the concentration from the 1st stage (the advective transport in the same time step).

3rd stage. We correct the concentration in every part, using the result from the dispersion part of the transport. This can be done by assigning the concentration in the mesh point to all particles in its neighbourhood. The solution in one time step is hereby finished and we can continue with the next time step.

The method of characteristic curves is suitable for problems with a large Peclet number since there are no oscillations and it does not suffer so much from the numerical dispersion.

Instead of these problems, other difficulties with particles occur in this method. Above all, the precision of the method depends not only on the density of the net but also on the number of particles. It is advisable (see Konikow and Bredehoeft 1978) to use from four to nine particles per block. One has to keep an eye on the number of particles during the calculation because particles can sometimes leave the domain. It is necessary to add particles at the infiltration boundary. One has to make sure that particles cannot leave the domain through the impermeable boundaries. The number of particles is large and that causes high time requirements of the calculation. Determining in which neighbourhood the given particle is and determining the velocities of all the particles cause the highest time requirements.

We have to realise that the particles used in this method are only a means of the advective transport's solution and they do not represent particles of the pollutant. This method is almost exclusively combined with the finite differences method, which solves the dispersion part of the transport.

6.1.5
Random Walk Method

This method belongs among the particle methods as well as the previous one. A drift of particles is caused by the groundwater flow, but this is where the similarities between the two methods end. In the random walk method the particles directly represent the pollutant's quantity (every particle has the same given quantity of the pollutant). This given quantity does not change in time if we do not consider a decay. Modelling pure advective transport is easy, we just have to let the particles be drifted by groundwater.

Some older works (such as Ito 1951) deal with dispersive transport and these show that the Fokker-Planck equation

$$\frac{\partial f}{\partial t} = \frac{\partial^2}{\partial x_i^2}(D_{ii}f) + \frac{\partial^2}{\partial x_i \partial x_j}(D_{ij}f) - v_i \frac{\partial f}{\partial x_j} \tag{6.35}$$

has a solution that can be explained as the distribution of a large number of particles. The position of every particle in the domain can be determined as (see Ackerer 1988)

$$x_i^c(t + \Delta t) = x_i^c(t) + v_i \Delta t + Z_{ij}\sqrt{2D_{ij}\Delta t}, \tag{6.36}$$

where x^c are coordinates of a particle at times t and $t + \Delta t$. Δt is the time step and Z_{ij} are random numbers governed by the normal distribution. The first term on the right side of Eq. (6.36) stands for the preceding position of the particle, the second is the advective term and the third is the term that describes a disperse behaviour of particles. The influence of dispersion is to blame for the random distribution of particles around their average position. They are distributed by the normal distribution. The average position is given by the advection. The variance of the distribution is determined by

$$\sigma_{ij}^2 = 2D_{ij}t. \tag{6.37}$$

The governing equation of the dispersion transport of pollutants without sorption has this general form

$$\frac{\partial C}{\partial t} = \frac{\partial}{\partial x_j}\left(D_{ij}\frac{\partial C}{\partial x_j}\right) - \frac{\partial}{\partial x_i}(v_{pi}C). \tag{6.38}$$

The Fokker-Planck equation [Eq. (6.35)] is identical with this equation only if the dispersion coefficients are constant. As they depend on the velocity of the flow, it is difficult to fulfil this assumption. We need to introduce a velocity reduction which is used to determine the advection (see Kinzelbach 1988)

$$v_i' = v_i + \frac{\partial D_{ij}}{\partial x_j}. \tag{6.39}$$

An advantage of the random walk method is an easy algorithm and a transparent approach. Additionally, this method is suitable even for a high Peclet number and it can tackle pure advection without any problems (this case is the most suitable because there is no need to generate random numbers). Somewhat controversial is another advantage of the random walk method, the non-existence of the numerical dispersion (see the introduction to this chapter). This is true only if we want the positions of particles in the domain, whereas if we want to calculate the concentration of pollutant in groundwater (which is the most common requirement), we have to use a secondary net to count the particles. Then we can reconvert the mass of the pollutant per water volume in one cell of the net. This procedure introduces the numerical dispersion but it is significantly smaller than by other methods. The larger the size of the cell, the larger is the numerical dispersion. On the other hand, the size of the cell cannot be decreased arbitrarily because a statistical method cannot cope with an infinitesimal size. According to recommendations from different authors (for example Kinzelbach 1988) there should be at least 20 particles in one cell of the net.

This method has some disadvantages, however. The most common boundary condition (constant values of concentration) is not suitable because it leads to the requirement of continuous addition of new particles into the domain. This is overcome by the presumption that the new particles move along the same trajectories as did the previous particles. Now we can add together all the particles in every time step and the result will correspond to the boundary condition. Another problem is caused by the boundary condition that presumes no outflow of pollutants from the domain. In this case, we have to stop the particles on the border of a domain, and this is closely connected with problems in software development.

The other disadvantage is the time requirements of the algorithm, as precise results need a large number of particles. The greater the dispersion coefficient, the larger has to be the number of particles. Very important is the method used to determine particles' velocity in each time step. Thus, the random walk method is always used together with another method for groundwater flow modelling which determines velocity. An often-used method is the finite differences method (see Kinzelbach 1988). In this case we have to find out the position (in which cell) of every particle in each time step. When using a large number of particles, this is a time-requiring process. The same goes for the finite element method. In addition, it is obvious that determining the velocity of the flow in a given point can cause problems. The finite element method can serve as an example. In the case of an often-used triangular element the velocity is constant in the whole element. All the particles inside have the same velocity. This leads imprecision in the transport's solution.

The most effective method to determine velocity is the boundary element method. This method enables an explicit calculation of velocities in an arbitrary point inside a domain. The velocity is continuous and none of the aforementioned problems occurs. This combination is used by the BEFLOW system.

6.2
Equilibrium Sorption

We analysed different kinds of sorption in Chapter 3.1.3 and 3.1.5. Here, we focus on how the methods can cope with the equilibrium sorption. As mentioned in Chap. 3, the simplest case of a sorption is a linear isotherm where the amount of the pollutant sorbed on the soil complex is in a linear relationship with the pollutant's concentration

$$S = K_d C. \tag{6.40}$$

K_d is the so-called distribution coefficient. The concentration S is a nondimensional quantity because it stands for a proportion between the mass of the sorbed pollutant and the mass of the skeleton. The concentration C has a dimension (ML^{-3}), for which reason the coefficient of distribution has the dimension $(M^{-1}L^3)$.

6.2.1
Analytic Solution

Let us return to the analytic solution from the Section 6.1.1. We will once again consider a semiindefinite band parallel with the x-axis. Let us have only one constant value of the velocity of groundwater flow in pores v_p in the whole domain. The effects of sorption are caused by a constant retardation factor R. The basic equation has this form

$$R\frac{\partial C}{\partial t} = D\frac{\partial^2 C}{\partial x^2} - v_p\frac{\partial C}{\partial x} - \lambda RC, \tag{6.41}$$

where λ is a constant of decay for a radioactive pollutant and the retardation factor R is defined as

$$R = 1 + \frac{\rho}{m_e}K_d. \tag{6.42}$$

This is a solution of the linear isotherm. The solution of this equation with the same boundary and initial conditions as in Section 6.1.1 is (see van Genuchten 1981)

$$\frac{C(x,t)}{C_0} = \frac{1}{2}e^{\left(\frac{(v_p-U)x}{2D}\right)}\text{erfc}\left[\frac{Rx-Ut}{2\sqrt{DRt}}\right] + \frac{1}{2}e^{\left(\frac{(v_p+U)x}{2D}\right)}\text{erfc}\left[\frac{Rx+Ut}{2\sqrt{DRt}}\right], \tag{6.43}$$

where U is a symbol for

$$U = \sqrt{v_p^2 + 4DR\lambda}. \tag{6.44}$$

6.2.2
Finite Element Method

This method can easily deal with a linear sorption. The basic equation [Eq. (6.11)] stays without a change. The only change occurs in the relations which determine the matrices. Relations (6.13) of the Petrov-Galerkin method change to

$$H_{ij} = \int_\Omega \left(D_{km}\frac{\partial N_i}{\partial x_k}\frac{\partial W_j}{\partial x_m} - v_k\frac{\partial N_i}{\partial x_k}W_j - \lambda RN_iW_j\right)d\Omega$$

$$R_i = -\int_\Gamma v_m D_{km} \frac{\partial N_i}{\partial x_m} W_j d\Gamma \tag{6.45}$$

$$M_{ij} = -\int_\Omega RN_i W_j d\Omega,$$

where N_i and W_j are the base and the weight functions, respectively. If we want to use another isotherm (not linear), Eq. (6.11) ceases to be linear. The non-linearity is solved inside the time steps. This means that we use the concentration from the previous time step to calculate the retardation factor and we do not solve the non-linear equation. One can see this in the UNSDIS program (see Chap. 9.2).

6.2.3
Particle Methods

We will take the random walk method as an example. The sorption appears in the reduction of porous velocity and in the reduction of the dispersion coefficient since we can derive Eq. (6.41) using the retardation factor, and thereafter we have

$$\frac{\partial C}{\partial t} = \frac{\partial}{\partial x_i}\left(\frac{D_{ij}}{R}\frac{\partial C}{\partial x_j}\right) - \frac{\partial}{\partial x_i}\left(\frac{v_{Pi}}{R}C\right) - \lambda C. \tag{6.46}$$

The sorption is present in this method in a term containing the velocity of the groundwater flow, the dispersion coefficient and the retardation factor. The solution will be identical with the one without sorption (Chap. 6.1.5). There is a term in Eq. (6.46) that simulates the pollutant's decay. There are two different approaches to this case. The first (see Prickett et al. 1981) randomly decreases the number of particles with a probability

$$p = \lambda \Delta t. \tag{6.47}$$

Δt time step has to be chosen sufficiently small to keep the probability p smaller than 1. This method was abandoned because it required a large number of particles to prevent a radical decrease in precision. The second approach uses a decay of mass directly on each particle. The initial mass M_p carried by one particle decreases as follows

$$M_p(t) = M_p(t_0)e^{-\lambda t}. \tag{6.48}$$

When we use a non-linear isotherm, the retardation factor is a function of concentration and relation (6.46) is non-linear. To solve this problem, we have to set the concentration in every time step. To do this we need a secondary grid. We have a different retardation factor in each cell of the net for a given time step. Then we have to correct the velocities and the dispersion coefficient by this factor.

6.3
Non-Equilibrium Sorption

By non-equilibrium sorption we consider even the rate of sorption. One governing equation is replaced by a system of two equations (see Chap. 3.1.5). The simplest case is again the linear sorption in the saturated zone, which is described by these equations

$$\frac{\partial}{\partial t}\left(\frac{\rho}{m_e}S+C\right) = \frac{\partial}{\partial x_i}\left(D_{ij}\frac{\partial C}{\partial x_j}\right) - \frac{\partial}{\partial x_i}(v_{pi}C) - \lambda\left(\frac{\rho}{m_e}S+C\right)$$

$$\frac{\partial S}{\partial t} = \alpha K_d C - S(\alpha - \lambda),$$
(6.49)

where C is a concentration of the pollutant in the solution, S is a concentration of the pollutant sorbed on the soil skeleton, ρ is a bulk density of grains of the skeleton, m_e is the effective porosity, λ is the decay rate and α is a coefficient of the rate of sorption. In contrast to the previous case, the solution cannot be transformed only to one unknown parameter with the help of the retardation factor, but we have two unknown concentrations in every time step.

The first equation is derived for further solution by substituting for the sorbed pollutant's derivative with respect to time from the second equation.

$$\frac{\partial C}{\partial t} = \frac{\partial}{\partial x_i}\left(D_{ij}\frac{\partial C}{\partial x_j}\right) - \frac{\partial}{\partial x_i}(v_{pi}C) - \lambda\left(\frac{\rho}{m_e}S+C\right)$$

$$- \alpha K_d \frac{\rho}{m_e}C + \frac{\rho}{m_e}S(\alpha - \lambda).$$
(6.50)

Then the first equation is used to calculate the concentration C and the second for the concentrations S.

6.3.1
Finite Element Method

We use the finite element method to solve the first equation from system (6.49) and the second equation is solved by the finite differences method. From the first equation we have a basic equation formally analogous to Eq. (6.11). The matrices from Eq. (6.11) can be, for the Petrov-Galerkin method, written as

$$H_{ij} = \int_\Omega \left[D_{km}\frac{\partial N_i}{\partial x_k}\frac{\partial W_j}{\partial x_m} - v_k\frac{\partial N_i}{\partial x_k}W_j - \left(\lambda + \alpha K_d\frac{\rho}{m_e}\right)N_i W_j\right]d\Omega$$

$$R_i = -\int_\Gamma v_m D_{km}\frac{\partial N_i}{\partial x_m}W_j d\Gamma + \int_\Omega (\alpha - 2\lambda)\frac{\rho}{m_e}S_i W_j d\Omega$$
(6.51)

$$M_{ij} = -\int_\Omega N_i W_j d\Omega,$$

where the concentrations S_i in every mesh point in a given time step are considered known from the previous time step. To solve the second equation from system (6.49), we have to represent the courses values of S and C in a time step similar to Eq. (6.10)

$$S = (1-\xi)S_1 + \xi S_2 \quad \xi \in \langle 0,1 \rangle. \tag{6.52}$$

If we substitute a difference for the time derivative to Eq. (6.49), we have

$$\frac{S_2 - S_1}{\Delta t} = \alpha K_d[(1-\xi)C_1 + C_2] - [(1-\xi)S_1 + \xi S_2](\lambda - \alpha) \tag{6.53}$$

after derivations, we gain

$$\left[\frac{1}{\Delta t} + \xi(\lambda - \alpha)\right]S_2 = \left[\frac{1}{\Delta t} - (1-\xi)(\lambda - \alpha)\right]S_1 + \alpha K_d[(1-\xi)C_1 + \xi C_2]. \tag{6.54}$$

This equation yields the values of concentration S in mesh points in a given time step.

6.3.2
Random Walk Method

The solution for non-equilibrium sorption stems, in the random walk method, from the procedure that (Kinzelbach 1988) used on the transport of pollutants in soils with two different values of porosity.

Let us divide the particles into two groups. In the first group there are particles of the solution and in the second there are the sorbed particles. To correctly divide the particles into these two groups we define these probabilities as p_1 and p_2.

$$p_1 = \alpha \Delta t K_d \frac{\rho}{m_e} \quad p_2 = \alpha \Delta t, \tag{6.55}$$

where p_1 is the particle's probability of being from the 1st group and p_2 is the particle's probability of being from the 2nd group. We generate a random number X using the normal distribution in $\langle 0,1 \rangle$ interval for every particle in each time step. If $X < p_1$, particles from the 1st group are transferred to the 2nd group in the next step. If $X < p_2$, particles from the 2nd group are transferred to the 1st group in the next step. Δt time step must be chosen so that p_1, p_2 are smaller than 1. The new position of particles in the 1st group is set by Eq. (6.36) and the position of the particles in the 2nd group does not change.

References

Abramovitz M, Stegun IA (1972) Handbook of mathematical functions. Dover Publishers Inc., New York

Ackerer Ph (1988) Random-walk method to simulate pollutant transport in alluvial aquifer or fractured rocks. In: Custodio E, et al. (eds) Groundwater Flow and Quality Modelling. D. Reidel, Dordrecht, pp. 475–486

Bear J (1972) Dynamics of fluids in porous media. American Elsevier, New York

Huyakorn PS, Nilkuha K (1979) Solution of transient transport equation using an upstream finite element scheme. Apll. Math. Model., 3:7–17

Ito K (1951) On stochastic differential equations. Am. Math. Soc., New York

Kinzelbach W (1988) The random walk method in pollutant transport simulation. In: Custodio E, et al. (eds) Groundwater Flow and Quality Modelling. D. Reidel, Dordrecht, pp. 227–245

Konikow LF, Bredehoeft JD (1978) Computer model of two-dimensional solute transport and dispersion in ground water: U.S. Geological survey techniques of water-resources investigations. Book 7, Chap. C2, 90 p.

Kovarik K (1978) Resení hydrodynamické disperze metodou konecných prvku. Vodohosp. cas. 1978, vol. 26, str. 204–216

Ogata A, Banks RB (1961) A solution of the differential equation of longitudinal dispersion in porous media, USGS Prof. Paper 411-A

Prickett TA, Naymik TG, Lonnquist CG (1981) A random walk solute transport model for selected groundwater quality evaluations. Illinois State Water Survey, Bulletin 65, p. 103

Reddel DL, Sunada DK (1970) Numerical simulation of dispersion in groundwater aquifers. Hydrology papers, Colorado State University, 41

van Genuchten MT (1981) Non-equilibrium transport parameters from miscible displacement experiments. Research Report No. 119, U.S. Salinity Lab., USDA

7
Comparison of Properties of All the Methods

This chapter is a review of the important properties of every method listed in Chapters 5 and 6. It is aimed to help with the choice of the particular method used in the model. Basically, there are two different aspects to each method. The first is the point of view of a hydrogeologist trying to solve a given problem, and the second is that of the software system developer who prepares the model.

7.1
Groundwater Flow Models

This section focuses on the comparison of properties of methods used in numerical solutions of a groundwater flow in the saturated zone (see Chap. 5). First, we will focus on the properties important for the geologist when preparing data for a mathematical model.

- *Fit of the domain's geometric shape.* First, one has to choose the domain of the problem's solution. The choice is mostly restricted by the following condition: the domain must have given boundary conditions (e.g. a geological interface, a river etc.) because the model has to be set up from known coordinates of the points. These coordinates are nowadays obtained from digitised maps. At this stage the finite differences method differs from all the other methods. The principles of this method imply the use of a regular net, mostly a rectangular one. This method requires the greatest simplification by the input of the domain's shape. This disadvantage caused it to be no longer used in problems which require a high precision of input. On the other hand, large-scale geological models are not so sensitive to the shape of the domain and the finite differences method can still be used here. The method is unsuitable in cases when the domain's shape matters (such as remediation).
- *Fit of the boundary conditions.* This criterion is a copy of the previous one as the boundary of the domain is often the place where a boundary condition is given. Here, the finite differences method is the least precise.
- *Non-homogeneous media.* Hereby we understand the ability of the method to match a variety of geological properties and the change of a coefficient of hydraulic conductivity. If we judged the methods with respect to this criterion, the finite element method would be clearly in the lead. Here, one can give different values of the coefficients in each element (this is allowed by

the finite differences method as well) and some types of elements enable changing the coefficient's value even inside every element. This criterion is not very well met by the BEM or DRM methods, since they can only solve problems in partially homogenous domains (or domains where the coefficient of hydraulic conductivity changes according to a known function, which is not the case with the groundwater flow models).

- *Introduction of point sources.* It is often necessary to introduce a point source to a model of groundwater flow. This is mostly a well (or a borehole) with a given yield. In this case, BEM and DRM are the most suitable methods because they allow the source to be placed arbitrarily inside the domain without having to change the net. The determined values of the potential in the point source and in its neighbourhood are very precise. The FEM and the finite differences methods, on the other hand, require that the source is placed in a net's node and the precision of the results depends on the net's density in the neighbourhood of the point source. If we want a model to check various positions of the source (e.g. remediation, pumping), it saves a lot of time when we do not have to change the net.
- *Match of a planar recharge.* This is the case of a recharge caused by rainfall or similar cases. This leads to problems only with BEM because one has to set up a new net that covers the whole domain (not only the domain boundary). This new net serves only to calculate the integrals that appear in the basic equation due to the planar recharge (see Chap. 5). This is true only for a planar model of the flow; in a 3-D case the recharge from rainfall can be dealt with without problems in BEM as well because it is a recharge through the boundary and not directly into the domain.
- *Balance.* We are talking about the balance of discharge and recharge in the entire domain. The balance is a great difficulty with FEM. The discharge through one side of an element can be easily calculated, but when using a polynomial of a lower order as a base function, the discharge is not a continuous function. This is true even for commonly used isoparametric elements. The balance can be set up by means of BEM and DRM because the result includes an exterior normal derivative in every element and this is used to determine the discharge through each element. The balance can be similarly calculated for each block in the finite differences method.
- *Velocity of flow.* In all the methods the problem of determining velocity is only secondary. Velocity is determined from the known values of potential after solving the basic equations. The velocity value is very useful in pollution transport models. The BEM and DRM methods fulfil this criterion best, as both determine the velocity using formulas for an arbitrary point inside the domain which is represented by its coordinates. The remaining two methods use an already existing net to calculate the velocities. The worst are the finite differences method and the FEM with the simplest linear element that can set only an average value of the velocity in a block (or an element). FEM, with base functions being polynomials of a higher order, can determine velocity more precisely.

- *Effort by data input.* At present, when computers and their software are developing so quickly, this criterion is not of greate importance. There are special professional programs that enable an easy input of data and a visualisation of the result. More than all other methods, FEM relies on the existence of these programs, as the amount of data is very large. This was the reason that restricted the use of FEM in the beginnings of modelling. BEM does not require such amounts of data and the preparation is therefore easier. In the case of a 3-D model, the amount increases rapidly and it nears the amount needed in FEM. The finite differences method takes advantage of the net's regularity and the amount of data is the smallest of all and so their preparation is the easiest.

These criteria are listed in Table 7.1. As mentioned above, there are two aspects to judging the numerical methods. The second is the point of view of the author of the software system. You will find a further comparison of the methods (now concerning the 2nd aspect) below.

Table 7.1. Comparison of numerical methods in the groundwater flow modelling (properties important for users)

Compared properties	Method			
	FDM	FEM	BEM	DRM
Match of geometric shape	Rough match	Good match	Good match	Good match
Boundary condition	Rough match	Good match	Good match	Good match
Non-homogeneous domains	Good match	Good match	Only partially homogeneous domains	Only partially homogeneous domains
Point sources	Problems, increased density of the net, complementary potential	Problems, increased density of the net	Good match	Good match
Planar recharge	Good match	Good match	Problems, a planar net required	Good match
Balance	No problems	Problems	No problems	No problems
Velocity of flow	Rough match	Depends on type of an element	Good match	Good match
Effort required by data input	Relatively low	High, special software needed	Relatively low	Relatively low

- *Type of system of equations.* As we have already seen in Chapter 4, all the compared methods are based on the weighted residuals method, which implies that their foundation is the solution of a system of linear equations. The finite differences and element methods use systems with symmetrical and banded matrices. A banded matrix is one with a narrow band of non-zero terms. This reduces the memory requirements of the solution and saves time. On the other hand, BEM and DRM result in a full non-symmetrical matrix. The lower number of equations compensates for the higher requirements (see below).
- *Number of equations.* This criterion is closely connected with the previous one. From the properties of BEM it follows that it is sure to have the lowest number of equations. DRM has only a slightly higher number of equations, but both the methods compensate for the advantage by a full and non-symmetrical matrix.
- *Optimisation of numbering.* Banded matrices often need a special numbering of variables to minimise the bandwidth of the system of equations. It is often claimed that the time requirement of the solution is in a linear relationship with the number of equations and in a quadratic relationship with the bandwidth. The optimising of the numbering is easy when a regular orthogonal net is used, as in the finite differences method. At first, one should number the variables in the direction where there is the lowest number of variables. To optimise the numbering of large and irregular nets that are found in FEM, we have to use some more complicated optimising algorithms such as the Cuthill-McKey algorithm. The algorithms are very complicated and often directly connected with the solution of the system of equations (e.g. the frontal solution algorithm used for the isoparametric elements). It

Table 7.2. Comparison of numerical methods in the groundwater flow modelling (properties important for software developers)

Compared properties	Method			
	FDM	FEM	BEM	DRM
System of equations	Banded, symmetrical	Banded, symmetrical	Full, unsymm.	Full, unsymm.
Number of equations	Very high	High	Low	Low
Numbering optimisation	Simple	Necessary and complicated	Unnecessary	Unnecessary
Calculation of local matrices	Simple	Depends on type of an element	Complicated even for simple elements	Complicated even for simple elements

is obvious that systems without a banded matrix do not need such algorithms.
- *Calculation of local matrices.* We showed in Chapter 5 that most methods use local and global matrices. By a local matrix we understand a matrix for every element that is added together to obtain the final matrix of the system (a global matrix). In the finite differences method the global matrix is set up directly and the setup of the matrix is the simplest of all the methods. The local matrix of FEM is a result of integration and the calculation depends on the type of element. The integration can be done analytically in the case of triangular elements with linear polynomials as base functions and the calculation does not require much time. More complex elements (e.g. isoparametric) require numerical integration. The analytic integration can be done in BEM and DRM only for the simplest element, but mostly the numerical integration or the seminumerical integration is needed.

7.2
Models of Transport of Pollution

In this part we will summarise the properties of pollution transport models that we discussed in Chapter 6. We compare four methods (see Table 7.3). The first two methods, FEM and the dual reciprocity method, belong to the direct methods that deal with the influence of the advective term directly. The remaining two methods represent the particle methods. These need another method that serves to calculate the disperse transport (in the case of the method of characteristic curves) or to determine the velocity of the flow at the coordinates of the particles (in the case of the random walk method). The most common method is the finite differences method, although in case of the random walk method BEM is the most suitable (see Chap. 9). Here, we concentrate on the ability of methods to handle problems characteristic for pollution transport.

- *Oscillations.* This phenomenon is characteristic for the direct methods. Oscillations appear in the proximity of the transition zone. The more dom-

Table 7.3. Comparison of numerical methods used in the transport of pollution

Compared properties	Methods			
	FEM	DRM	MOC	Random walk
Oscillations	Appear	Appear	None	None
Numerical dispersion	Appears	Appears	Only restricted	Only restricted
Pure advection	Impossible to solve	Impossible to solve	No problems	No problems
Boun. con. 1st kind	No problems	No problems	No problems	Problems

inant is advective transport, the narrower is the transition zone and the oscillations are larger.
- *Numerical dispersion.* Hereby we understand an artificial spread of pollution into an area larger than the area supposed in the real situation. It is caused by dividing the domain into elements (or blocks) and by interpolating the concentration function in the net. In order to remove this phenomenon, the direct methods use different modifications (e.g. Petrov-Galerkin method, see Chap. 6.2), but generally, the more one tries to remove the numerical dispersion the greater the oscillations that appear in the solution. The particle methods do not suffer from numerical dispersion so significantly. The random walk method shows no sign of numerical dispersion as long as we are in a particle cloud. If we want to redetermine the pollutant concentration from these particles, we have to use an additional net and count the particles in its cells. This introduces the numerical dispersion. This result is not applied further in the following steps, thus the influence of the numerical dispersion is negligeable. A non-existing or limited numerical dispersion is a valid argument in favour of the particle methods.
- *Pure advective transport.* The problem where there is no dispersion and the transport is only advective does not appear in porous media. It is more typical for a transport of pollution in fluid (pollution of rivers) where the dispersion is represented only by a molecular diffusion which is neglectable comparing to the velocity of flow. The direct methods cannot solve this case; only the particle method can deal with this type of transport.
- *Setting the boundary conditions.* No problems occur in the direct methods. The only method that has some problems with the boundary condition of the 1st kind is the random walk method. It has to permanently emit new particles. To solve this, we use the particles from the previous time step (see Chap. 6).

8
Examples of the Use of Models in Practice

This chapter presents examples of practical applications of all the methods mentioned in the previous chapters. Most applications listed here come from Slovakia or the Czech Republic, and they originated between the years 1990 and 1999.

8.1
Models to Determine Groundwater Resources

This is the oldest group of models. The mathematical models were first mainly used for these tasks but as the hydrogeological tasks changed, so also the ways in which the models were applied. Nowadays, models are used to assess the influence of buildings on groundwater or to assess variants of groundwater remediation.

8.2
Models to Assess Different Influences on Groundwater

These models assess the influence of dam constructions or drawdowns caused by other sources etc.

8.2.1
Example of Zilina Locality, Slovakia

This model was to predicted the groundwater level and the velocity of groundwater flow after filling the Zilina reservoir (see Kovarik 1994). The problem obviously requires the use of a 3-D model. Here, we solved a steady flow by means of a software system FE3D which uses the finite element method with subparametric elements of 20 nodes (see Chap. 5.1.3.3).

8.2.1.1
Geological Structure of Locality

A large geological survey shows that the locality of the reservoir consists of rocks from the Central-Carpathian Paleogene and Quaternary.

The Hricov-Podhradsky paleogenous lithological facies of flysch series of strata consists of sandstone, siltstone and claystone with calcareous cement. In the area of interest the maximal thickness of a continuous layer of sandstones is 7 m.

In Zilina valley the series of paleogenous strata is generally tilted to the north with an angle from 10° to 20°. In the neighbourhood of the dam construction, some layers were detected which were tilted to the south or southeast with an angle of 10° to 15°. The Paleogene is strongly weathered to a depth of 0.5 to 3 m under its surface. The weathered zone reaches a depth of 4 to 10 m and some irregularities reach even 20 to 25 m. We can assume a faulted zone within reach of the weathering with dislocations of 1 m width. Here the weathering reaches even deeper.

The Quaternary layer in the area of the model consists of fluvial alluvions of loam, sand and gravel with a thickness of 2 to 5.5 m. The gravel is composed of pebbles averaging from 5 to 10 cm. The volume of the pebbles is almost 60–70%. The gravel fillings are loamy sand. The coefficient of hydraulic conductivity of the loamy sand was set by pumping tests to $1.5 \times 10^{-4}\,\mathrm{m\,s^{-1}}$ and the same coefficient of the sandy gravel is $4.5 \times 10^{-4}\,\mathrm{m\,s^{-1}}$.

The coefficients of hydraulic conductivity in the model were set using the previous modelling works or water pressure tests.

The previous modelling works for Quaternary (models of assessing the function of biocorridor) provided an average coefficient of hydraulic conductivity in the horizontal direction ($k_x = 4.5 \times 10^{-4}\,\mathrm{m\,s^{-1}}$). The sandy gravel has an estimated anisotropy of 9 and so the vertical coefficient of hydraulic conductivity is $k_z = 5 \times 10^{-5}\,\mathrm{m\,s^{-1}}$. Near the Vah terrace the horizontal coefficient of hydraulic conductivity drops to $5 \times 10^{-5}\,\mathrm{m\,s^{-1}}$.

The coefficient of hydraulic conductivity for the weathered zone was acquired from reviewed water pressure tests and its value was set to $1.3 \times 10^{-5}\,\mathrm{m\,s^{-1}}$.

We reviewed the pressure tests by the same approach as above for the remaining series of strata and obtained values around $1 \times 10^{-6}\,\mathrm{m\,s^{-1}}$ that corresponded well with previous surveys.

8.2.1.2
Geometry of the Model

These data imply that there are three basic layers of the porous medium, namely Quaternary steps, a weathered neogeneous subsoil and deeper paleogeneous layers. These layers are almost horizontal in the flooded area but behind the dam profile they start to climb, so that their direction and the horizontal direction include approximately 10°. This position of layers was considered in the model setup. The best image of the model gives a 3-D view of the model shown in Fig. 8.1. The medium was divided into five layers. The data of every layer can be found in Table 8.1.

The numbering of layers starts from the lowest layer. The first three layers represent the paleogenous subsoil each with a different permeability, the

8.2 Models to Assess Different Influences on Groundwater 133

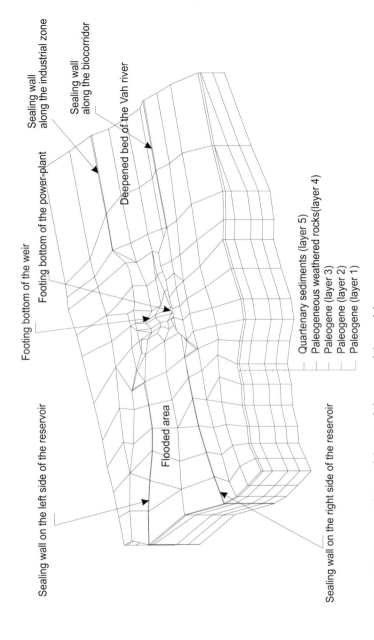

Fig. 8.1. Zilina, spatial view of the net of elements of the model

134   8 Examples of the Use of Models in Practice

Table 8.1. Coefficients of layers (Zilina dam)

Layer	Geology	Average thickness	K_x (m s^{-1})	K_z (m s^{-1})
1	Paleogene	20	5.7×10^{-7}	5.7×10^{-7}
2	Paleogene	20	1.2×10^{-6}	5.0×10^{-7}
3	Paleogene	20	1.0×10^{-6}	5.0×10^{-7}
4	Weathered Paleogene	24	1.2×10^{-5}	1.6×10^{-6}
5	Quaternary	4	4.5×10^{-4}	5.0×10^{-5}

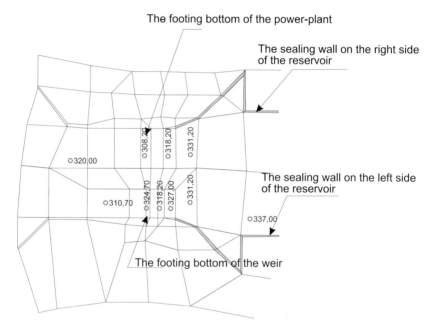

Fig. 8.2. Zilina, detail of the net near the dam footing bottom

fourth layer is the weathered surrounding rock and the last layer is the Quaternary step.

The dam construction is set into the situation described above and it is characterised by two parts of a footing bottom. The first part is the weir and the second the power plant. The part of the footing bottom which is covered by the concrete of the dam construction is presumed impermeable according to the project (see detail of the net in Fig. 8.2).

The flooded area is in front of the dam profile and is restricted by earth dams. Since the dams were sealed using a membrane that continuously changes to an underground sealing wall, and because the model should be a model of the subsoil, these dams were not included. The flooded area is here restricted by underground sealing walls (see Fig. 8.1) that penetrate only layers 5 and 4 (Quaternary sediments and weathered paleogenous rocks). According to the project, the coefficient of hydraulic conductivity should be $1 \times 10^{-7}\,\mathrm{m\,s^{-1}}$, which is the value given in elements of the sealing wall. The bottom of the flooded area is 337 m above sea level.

Under the dam profile the Vah river bed is deepened. It cuts into the paleogenous layers to a level of 323.3 m above sea level. There are sealing walls along both sides of the river bed. The left wall should seal off the eastern industrial zone and the right wall was designed to seal off the biocorridor. Both the walls have the same hydraulic conductivity coefficient, being $1 \times 10^{-7}\,\mathrm{m\,s^{-1}}$, and they penetrate only the Quaternary layer.

In the ground plan the model is limited by two boundary conditions, the existing Vah terrace on one side and the edge of the biocorridor on the other. The net has the same projection into the ground plan in all the layers because the sides of the elements are vertical.

Nodes of the net can be divided into three groups as follows:

- Vertices of elements in the 1st layer. These were directly digitised from maps.
- Vertices of elements in other layers. These were given by their altitudes and have the same x, y coordinates as the nodes from the previous group.
- Nodes in the middle of elements' edges. Their coordinates were generated by the program as an arithmetic average of the coordinates of corresponding vertices.

8.2.1.3
Solution Variants

After setting up the model, three variants were solved, all three founded on the same boundary conditions. The boundary condition of the 1st kind was given by a maximal water stage of 352 m above sea level in the flooded area and a minimal water stage of 325.4 m above sea level in the deepened river bed. Besides these boundary conditions we introduced a level in the biocorridor in the top layer (Quaternary), which changes with respect to the surface slope. Its setting stemmed from older modelling works on assessing the effects of the biocorridor, and the level was set between 336 and 338 m above sea level. There is the Rosina stream in the left-side alluvium which is one of the boundary conditions in the Vah Quaternary with an altitude of 335.5 m above sea level.

The first variant is based on the described geometry and on the boundary conditions. Boundaries of the flooded area are the right and left underground sealing walls (USW) and boundaries of the deepened river bed behind the dam profile are again the right and left USW. The equipotentials in the third layer and in the top layer which are a result of the model, are in Figs. 8.3 and 8.4, respectively.

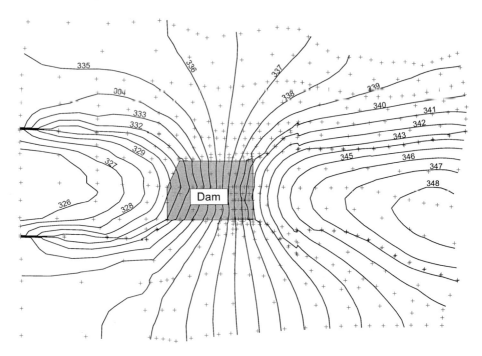

Fig. 8.3. Zilina, equipotentials in the 3rd layer, 1st stage

In contrast to the 1st variant, there are no sealing walls behind the dam profile in the second variant (this means in practice that they were not built as far as the dam profile, see Fig. 8.5).

The 3rd variant shows the hypothetical possibility of a decrease in the sealing effect of the USW in the flooded area but only in the area of intake wing walls in the 5th layer (Quaternary layer, see Fig. 8.6).

As the amount of velocity data is very high, it is suitable to do a statistical comparison for every layer and every variant. Obviously, the most interesting pieces of information are the maximal and minimal values of velocity in each layer (see Table 8.2).

This comparison shows that the effects of changes in the first layer can be ignored, as the velocities are influenced by these changes in the Quaternary only marginally. The greatest increase in velocity can logically be found in the last Quaternary layer in the 3rd variant, which simulates a leakage of the USW in elements near the corner of the USW. Here, the velocity reaches its maximum.

The effects of the sealing wall on the dam construction are only minimal, though their effects on the flow in the Quatenary layer are significant. This can be seen from a comparison of the velocity in elements that form the left and right Quaternary along the deepened river bed. If the walls were not there, the

8.2 Models to Assess Different Influences on Groundwater

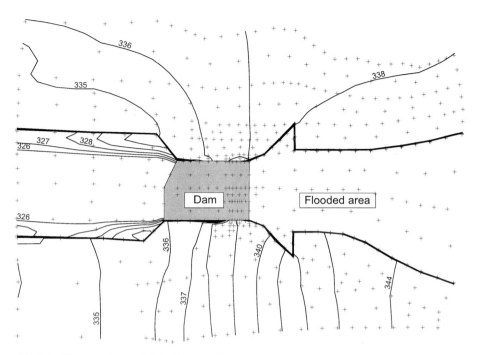

Fig. 8.4. Zilina, equipotentials in the upper layer, 1st stage

Table 8.2. Basic statistical data of velocities in layers (Zilina dam)

Stage	Type	Layer				
		1	2	3	4	5
1st	Minimum	4.98×10^{-9}	1.20×10^{-8}	1.01×10^{-8}	7.50×10^{-8}	7.77×10^{-8}
	Maximum	6.74×10^{-8}	2.65×10^{-7}	8.47×10^{-7}	1.43×10^{-5}	9.36×10^{-5}
	Mean	2.41×10^{-8}	6.31×10^{-8}	9.19×10^{-8}	7.95×10^{-7}	5.88×10^{-6}
2nd	Minimum	5.52×10^{-9}	1.38×10^{-8}	1.09×10^{-8}	7.86×10^{-8}	8.02×10^{-8}
	Maximum	3.99×10^{-8}	1.54×10^{-7}	1.99×10^{-7}	9.24×10^{-6}	1.49×10^{-4}
	Mean	2.36×10^{-8}	6.15×10^{-8}	8.09×10^{-8}	7.12×10^{-7}	9.00×10^{-6}
3rd	Minimum	2.22×10^{-9}	6.87×10^{-9}	8.13×10^{-9}	4.38×10^{-8}	6.43×10^{-8}
	Maximum	6.76×10^{-8}	2.68×10^{-7}	8.48×10^{-7}	1.41×10^{-5}	1.01×10^{-4}
	Mean	2.45×10^{-8}	6.37×10^{-8}	8.79×10^{-8}	8.01×10^{-7}	1.09×10^{-5}

rate of filtration would increase three times to almost 5×10^{-5} m s^{-1}. The biggest increase is, of course, where the sealing wall is missing but it is caused by the change of the coefficient of hydraulic conductivity. Bearing in mind that the tail river bed, due to the deepening, reaches the paleogenous subsoil, we can

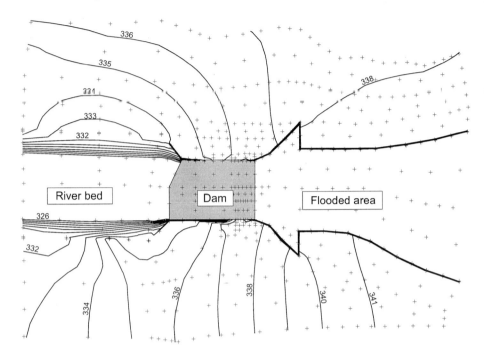

Fig. 8.5. Zilina, equipotentials in the upper layer, 2nd stage

say that the effects of the sealing wall behind the profile are insignificant and they influence only the Quatenary layer.

A totally different case are the effects of the USW in the flooded area. If there were a leak in the USW (3rd variant), the velocity would increase by one order as in the element 826 from 3.7×10^{-6} to 5.3×10^{-5} m s^{-1}. The element 826 is situated in the right dike (see Fig. 8.2) and this increase in velocity would probably cause the instability of the dike (a danger of suffusion or other danger). The 3rd variant simulates a leak of the USW in its sharp corner, where it joins the intake wing walls. This can be the most problematic place when building the USW. An obvious result is an increase in the groundwater level (e.g. for element 942, which is in the neighbourhood of the leak, the level increases from 338.23 to 341.15 m above sea level, which is the altitude of the terrain).

8.2.2
Example of Ziar Locality, Slovakia

This is a slightly untypical example as the aim of the model was to recommend variants of remediation in order to eliminate the danger of water source's pollution (Bansky and Kovarik 1995). At first sight, this model should belong to the next part of this Chap. about remediation models. It is not placed there

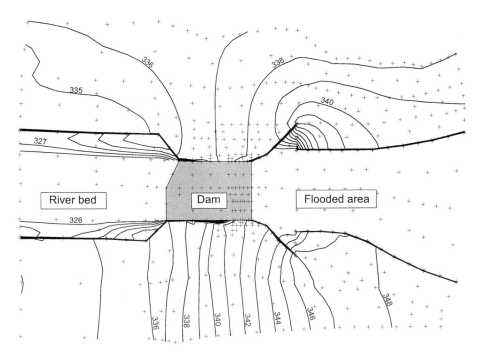

Fig. 8.6. Zilina, equipotentials in the upper layer, 3rd stage

because here only the modelling of the groundwater flow takes place and there is no pollution transport modelling. There is a source of drinking water in an industrial zone. Following the groundwater pollution survey performed in the entire industrial zone, there occurred a possibility of polluting the water source from local deposits of contaminated water inside the industrial zone. All this led to the decision to make different studies of the water source's remediation.

The water source is a catch drain. It is built near a factory along the Hron river bed, its distance from the river bed being 50–120 m. The length of the drain is 1074 m. The groundwater is collected by a stoneware drainage pipe with a diameter of 350 mm, and then deflected down by gravity into a collecting well Rotunda. From here, it is pumped to a water reservoir. The pumps are adjusted automatically according to the water level or manually from the vellum of a water treatment plant. The average water takeoff from the source is $32 \, \mathrm{l\, s^{-1}}$. The source was built according to a project from 1955. The collecting device is in a depth of 6–10 m under the terrain, i.e. almost in the neogenous subsoil.

The area of our interest is on the left side of the flood plain of the river which meandered in this area in the past. This influenced a settlement of different heterogeneous sediments with variable permeability. The aquifer consists of non-cohesive alluvial sandy gravel sediments which form a phreatic aquifer depending mainly on the water level fluctuations in the river and also on precipitation. River sediments of the river flat reach an average thickness of about 8 m and somewhere even 15 m. The collector of groundwater is composed of sandy gravel sediments with a thickness of 2 to 14 m. The Quaternary sediment's subsoil is formed by neogenous sediments. In the area of the model they are represented by an impermeable layer of tuffaceous rocks and at some places by semipermeable tuffaceous sands that occasionally transform into sandy clay.

We used BEM for the groundwater flow model. The whole domain was divided into 14 zones for geometric reasons (to match the shape of the takeoff drain) and for reasons connected with different filtration parameters. The division created 147 elements which include 81 boundary elements (with a set boundary condition). The rest were the interzone elements (separating the zones of unhomogeneities from each other). The coefficients of hydraulic conductivity were reviewed during the model's verification and the results range from 2.4×10^{-5} to 1×10^{-3} m s^{-1}. The catch drain was modelled using 38 elements, where 18 elements represent the segments between check wells S-1 to S-18 and two elements form the front of the drain. The arrangement of elements allowed us to determine the amount of water coming from the river and from the remaining alluvion.

An obvious boundary condition is the river. To weaken this boundary condition, we introduced a zone of clogged river banks and bed, whereby the condition was moved to the centre of the river bed. To set the water level in the river, we used a report by Bondarenkova (1993). This report was again referred to while setting the water surface slope of the Hron (it was 0.2%).

The first solution simulates the initial verified groundwater flow towards the catch drain. A total recharge to the catch drain is 32.78l s^{-1} which corresponds to the average takeoff from the source. A recharge from the river is 22.28l s^{-1} and from the remaining alluvion it is 10.5l s^{-1}. This is 32.03% of the source's yield. The results are published in form of a net of equipotentials and streamlines in Fig. 8.7.

8.2.2.1
Borehole Remediation

Building a hydraulic protection in form of a pumping well curtain belongs nowadays to standard remediation methods. The area of interest is not suitable for this method since it has gravel and sand layers of small thickness, implying a relatively small thickness of the aquifer. This restricts the yield of pumping wells, which means a small cone of depression. Additionally, the catch drain is long (more than 1000 m) and the number of pumping wells would have to be large. In the first proposal there were eight wells placed along the perimeter of the 1st protection zone of the source (close to the fence). These wells were designed for a maximal pumping yield of 0.8l s^{-1} per well. The pumping wells do not have to be elements of the net in the BEFLOW system and can be

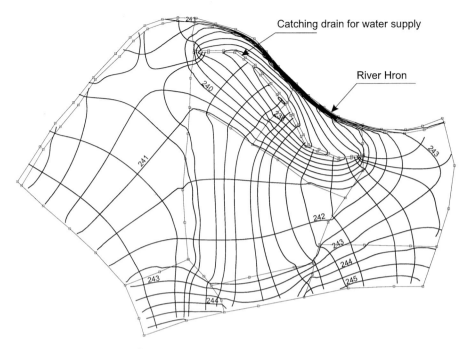

Fig. 8.7. Ziar, initial state, flow net

placed arbitrarily. The net of elements did not have to be changed after verification, we just had to add the pumping wells. The coefficients of hydraulic conductivity and the thickness of aquifer remained the same as in the verification model. The total yield of the hydraulic curtain is $6.4 \, \text{l s}^{-1}$. A recharge to the catch drain is $28.19 \, \text{l s}^{-1}$ where $22 \, \text{l s}^{-1}$ is a recharge from the river and $6.2 \, \text{l s}^{-1}$ was a recharge from the industrial zone. There are streamlines that penetrate the curtain and subsequently end up in the source. This means that the recommended curtain has a leak, which is represented by a non-zero recharge to the source from the industrial zone.

Hence, we added two additional wells to the hydraulic curtain in the 2nd variant, a well Z-9 between the wells Z-4 and Z-5 and a well Z-10 between the wells Z-5 and Z-6 (see the net in Fig. 8.8). We can again find places where the polluted water penetrates the curtain, as can be seen in the net of equipotentials and streamlines in Fig. 8.8.

8.2.2.2
Impermeable Wall Remediation

This type of remediation is one of the most common. It is an expensive way of remediation. This wall is in action permanently and cannot be turned on or

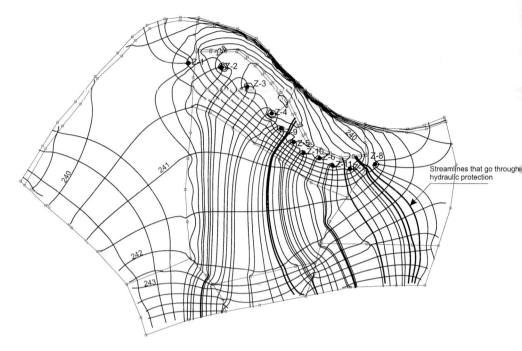

Fig. 8.8. Ziar, borehole remediation, flow net

off like the wells in the case of a suspected leakage of pollutant. This and the high cost are the main disadvantages. We placed the wall again close to the fence of the 1st protection zone of the source. The wall was assumed to be fully penetrated down to the impermeable subsoils. BEM requires a new net of elements and zones (see Fig. 8.9). The wall was presumed to be totally impermeable as an opening in the domain with a boundary condition of the 2nd kind $q = 0$. The building of the wall is an intervention in the groundwater flow beneath the factory. The main part of the groundwater flow is turned along the wall and it infiltrates into the river at the spot where the source ends (it can be clearly seen in the net of streamlines in Fig. 8.9). This remediation method presents another problem, the increase in groundwater level compared with the initial level. There is an average increase in level by 1.7 m (e.g. from 241.08 to 242.71 m above sea level). Moreover, the flow can partially penetrate the wall at its ends and some streamlines finally end up in the source. To ensure good protection, we have to place a short drain or two pumping wells near the ends.

The recharge to the source in this variant was decreased to $23.14 \, \text{l s}^{-1}$ and there is a recharge of $21.95 \, \text{l s}^{-1}$ from the river and only $1.19 \, \text{l s}^{-1}$ from the industrial zone (through the ends).

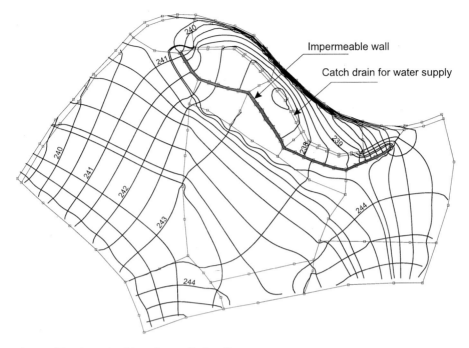

Fig. 8.9. Ziar, impermeable wall remediation, flow net

8.2.2.3
Drain Remediation

The third possibility to restrict the polluted water only to the industrial zone is the building of a remediation drain along the water source. The position of the drain is identical with the position of the wall in the previous paragraph. The altitude of the drain is the same as that of the source. The remediation drain is longer than the takeoff drain and has a length of 1210 m. It consists of five segments that are built so that they can be in action separately. They should be equipped by closures to separate the water takeoff of every segment. The recharge of the source in this case dropped to $20.6 \, \text{l s}^{-1}$. The recharge from the industrial zone ceased completely and all the $20.6 \, \text{l s}^{-1}$ comes from the river. The recharge of the remediation drain is $22.6 \, \text{l s}^{-1}$ which is caused by an increased recharge mainly at the south end of the drain. Recharges into each segment are listed in Table 8.3. In an effort to replace a decrease in the source's yield, we also judged variants with artificial small weirs so as to raise the water level in the river. These variants have numbers from 2 to 4. In variant 2, the water level in the river was raised by 1.3 to 242.5 m above sea level (see

144   8 Examples of the Use of Models in Practice

Fig. 8.10). Variant 3 presumed building a weir in the position of the collecting well Rotunda, raising the level in the river by 1.5 to 243.9 m above sea level. The last variant combined both cases.

One can see that the recharge of the remediation drain does not change and the recharge of the source does not reach the needed value (see Table 8.3).

The 5th variant stands for a possibility that the segment no.1 of the remediation drain is out of action as there is no pollution assumed in that area.

The mathematical model assessed effects of different methods of hydraulic protection of the source on its yield. The bigger variability aspect (possible

Table 8.3. Variants in drain remediation (Ziar)

No. of segment	Inflow to the remediation drain			
	1. variant	2. variant	3. variant	4. variant
1	3.298	3.333	3.299	3.452
2	1.108	1.077	1.100	1.109
3	5.009	5.042	5.11	5.31
4	6.139	6.14	6.235	6.492
5	7.038	7.038	7.272	7.482
Sum	22.6	22.63	23.016	23.845

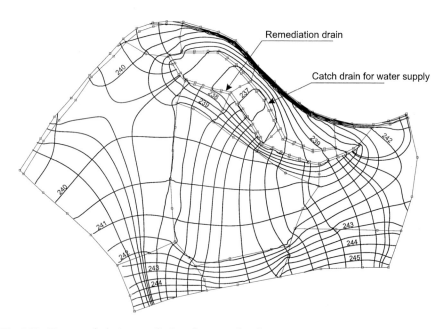

Fig. 8.10. Ziar, trench drain remediation, flow net of 2nd stage

Table 8.4. Comparison of all methods of remediation (Ziar)

		Without remediation	Method of remediation						
			8 wells	10 wells	Wall	Drain 1	Drain 2	Drain 3	Drain 4
Inflow to the well	From the river	22.3	22.0	21.9	21.9	20.6	22.2	23.0	24.4
	Polluted	10.5	6.2	5.2	1.2	0	0	0	0
	Whole	32.8	28.2	27.1	23.1	20.6	22.2	23.0	24.4
Remediation inflow		0	6.4	8.0	0	22.6	22.6	23.0	23.8

turning on and off according to need) is fulfilled only by borehole and drain remediation. Of these, borehole remediation cannot guarantee total isolation of the water source from the terrace and drain remediation has a tremendous impact on the source's yield that cannot be compensated for by an increase in the river's water level since the the hydraulic resistance is relatively high due to clogging and river regulation. We proposed using the water pumped from the wells or the drain as supply and service water needed for running the factory (see Table 8.4).

We listed this example here to show possibilities to assess different variants of protection of a water source without having to use any model of pollutant transport (using only a relatively simple model of groundwater flow). This model served as an initial study followed by further modelling.

8.2.3
Example of SNP Square, Bratislava, Slovakia

This model was part of the large project of completing the SNP Square in the centre of Bratislava and its aim was to assess the influence of underground objects on the groundwater level (garage, tunnels of a planned underground). We modelled the current state of the groundwater flow and how the groundwater flow would change after four construction stages that were defined by the project. The situation required a 3-D model due to the complexity of the construction and of the hydrogeological situation. This numerical model of a steady groundwater flow used BEM (see Drahos and Kovarik 1998a). As the entire report was too large to publish, we selected only the significant parts.

8.2.3.1
Setting up the Model

The model was set up for an approximately rectangular area based on the technical background provided by the project. The area of the model is only a part of a water-bearing rock continuum. The model was divided into three horizontal layers; the top layer simulates the Quatenary sediments consisting of gravel and sandy gravel as well as the concrete constructions. The middle layer was created to enable modelling of a flow under the concrete constructions in the pervious water-bearing gravel and in the thick neogenous series of water-bearing, less pervious strata below. The bottom layer consists of neogeneous sediments which, according to geophysical measurements, are hydraulically active down to 48 m below the terrain. Every layer is composed of triangular elements in order to match the shape of building constructions, horizontal geological boundaries and hydrogeological unhomogeneities. A north view of the net of triangular elements can be found in Fig. 8.11. This figure also shows a vertical division of the water-bearing rock in the model with the dominant thickness of the bottom neogenous layer. Figure 8.12 depicts the inner net of the vertical elements.

All three layers are split into numbered homogeneous zones with respect to the coefficient of hydraulic conductivity. There are 29 homogenous zones in the top layer (Fig. 8.13). Highly permeable Quaternary gravel is in zones 14, 15, 27, 28 and 29 with the coefficient $k = 2.1 \times 10^{-3}\,\mathrm{m\,s^{-1}}$. Less permeable terrace gravel ($k = 5 \times 10^{-4}\,\mathrm{m\,s^{-1}}$) occupies zones 9 to 13 and 22 to 25. Zone no.1 is the remainder of the water-bearing higher-terrace gravel that has a different value of the coefficient ($k = 4.1 \times 10^{-4}\,\mathrm{m\,s^{-1}}$). Zones 2 to 5 and 16 to 21 represent a sandy and loamy water-bearing deluvium of the neogenous sediments. Their coefficient of hydraulic conductivity is $k = 6.4 \times 10^{-6}\,\mathrm{m\,s^{-1}}$. The last three zones represent buildings with some parts of their construction below the groundwater level. Their coefficient of hydraulic conductivity was set to $k = 1 \times 10^{-15}\,\mathrm{m\,s^{-1}}$. In the four construction stages, other similar zones were added with the same permeability (all data above and below are for the current state).

The number of zones in the middle layer reaches 23 (Fig. 8.14). Highly permeable Quaternary gravel takes up zones 38, 39, 51 and 52 ($k = 5 \times 10^{-4}\,\mathrm{m\,s^{-1}}$). The neogenous sediments with the coefficient $k = 6.4 \times 10^{-6}\,\mathrm{m\,s^{-1}}$ occupy zones 30 to 32 and 40 to 44. Zones 33, 34 ($k = 1 \times 10^{-15}\,\mathrm{m\,s^{-1}}$) correspond to the only two buildings that reach down to this middle layer.

In the bottom layer there are only eight zones (Fig. 8.15). In zone 53 the Neogene has a hydraulic conductivity of $k = 1.3 \times 10^{-5}\,\mathrm{m\,s^{-1}}$ and in the remaining zones $k = 6.4 \times 10^{-6}\,\mathrm{m\,s^{-1}}$. No building reaches this layer in the current state.

All coefficients of hydraulic conductivity in all layers and zones were set using a hydrogeological survey. The vertical coefficients of hydraulic conductivity were assumed as $k_z = 0.1 k_x$ analogous to other similar problems.

Altitudes of the groundwater free surface served as altitudes of the top layer and can be seen in Fig. 8.16. We set a boundary condition of constant pressure in the bottom layer according to pressure heads measured in neogenous hydrogeological boreholes and according to the contours of potential. We also set a boundary condition of the 3rd kind on the sides (vertical) of the model that should simulate the fact that the area was only a part of a larger water-bearing layer.

8.2 Models to Assess Different Influences on Groundwater 147

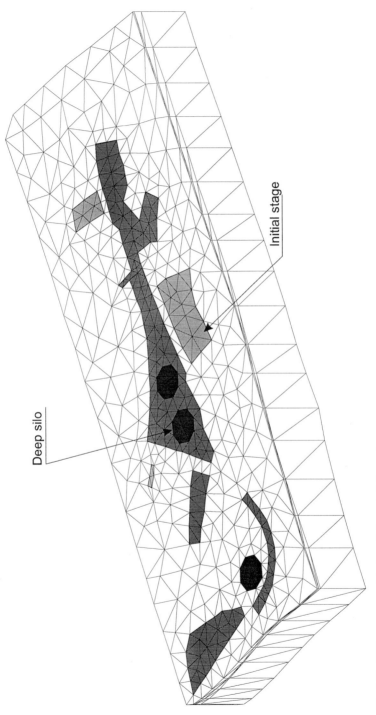

Fig. 8.11. SNP square, spatial view of the net of elements

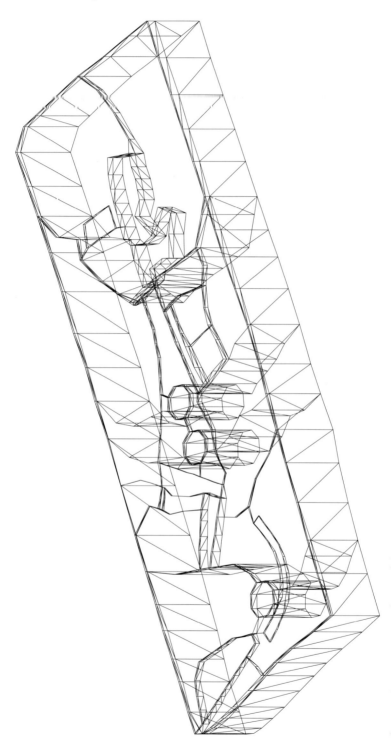

Fig. 8.12. SNP square, spatial view of the vertical elements

8.2 Models to Assess Different Influences on Groundwater   149

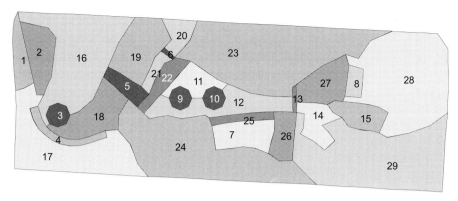

Fig. 8.13. SNP square, zones of unhomogeneities in the 1st layer

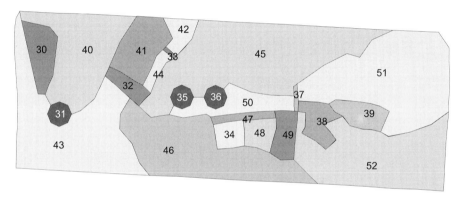

Fig. 8.14. SNP square, zones of unhomogeneities in the 2nd layer

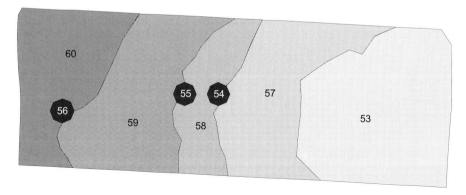

Fig. 8.15. SNP square, zones of unhomogeneities in the 3rd layer

150   8 Examples of the Use of Models in Practice

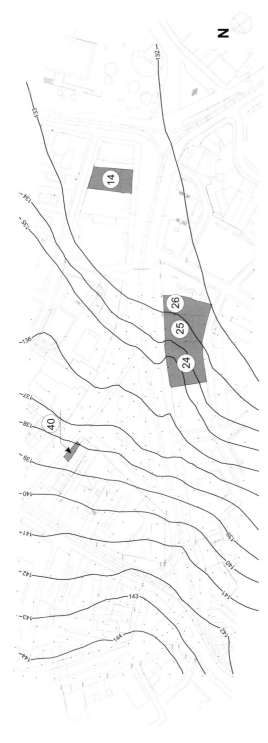

Fig. 8.16. SNP square, contour lines of groundwater level in the initial state

This stage pictures only a groundwater flow without influence from either the buildings that were to be constructed or the building activity itself. Even now there are buildings reaching below the groundwater level; building no. 40-a town house on Postova street 6; buildings no. 24-a bank (CSOB); no. 25 (Department store Dunaj)-a shopping centre; no. 26 (OTEX)-a shop and the building no. 14 (Theatre West). These objects are dark grey in Fig. 8.16.

Object no. 40 penetrates two layers (zones 6, 33) into an altitude of 137 m above sea level and the flow under this object can take place only in the neogenous subsoil. The same is true for object no. 26 (zones 7, 34) which has its basement at 129.2 m above sea level. The flow under objects nos. 24, 25 and 14 takes place in the Quaternary gravel and the buildings reach 131.7, 132.7, 131.5 m above sea level, respectively.

The described 3-D model was prepared for the current state of the groundwater free surface and was verified by a comparison of the predicted and the measured data. The determined groundwater levels (see Fig. 8.16) served as a prognosis of the change caused by the construction. The piesometric level in the bottom layer is higher than the level of the groundwater free surface and this implies a trend of a vertical water flow from the Neogene to the Quaternary. The differences in groundwater level between the top and the bottom layers are from 0 to 2 m.

8.2.3.2
Assessment of Construction Stages

Here we consider only the last, fourth, stage of construction. This stage represents the whole construction of three parking silos that reach the last neogenous layer, and a large underground parking with entrances that is situated under almost the entire area of the square. Moreover, further underground objects were added to the bottom side of the square. These objects are shown in Fig. 8.17 (grey colour).

The whole set of buildings is situated in the highly permeable Quaternary gravel and, therefore, we use an underground sealing wall with $k = 1 \times 10^{-9}\,\mathrm{m\,s^{-1}}$ for protection. In the top layer the hydraulic conductivity of the concrete is $k = 1 \times 10^{-15}\,\mathrm{m\,s^{-1}}$ (zones 14 and 15) and in the middle layer we set the coefficient of hydraulic conductivity according to the conductivity of the protection wall to $k = 1 \times 10^{-9}\,\mathrm{m\,s^{-1}}$. All the constructions are assumed to be at an altitude of 127.94 to 129.2 m above sea level which is the boundary of the Quaternary gravel and the neogeneous subsoil. The basement of the protection wall is in the less permeable neogenous sediments at an altitude of 120.4 m above sea level. The coefficient of hydraulic conductivity of the neogenous sediments is $k = 1.3 \times 10^{-5}\,\mathrm{m\,s^{-1}}$.

The modelled free surface of groundwater is shown in Fig. 8.17. The newly designed protection wall along with an existing barrier of object no. 14 (West Theatre) cause an elevation in groundwater level. This is also caused by a recharge of groundwater from the neogeneous subsoil of the side barriers. The

152   8 Examples of the Use of Models in Practice

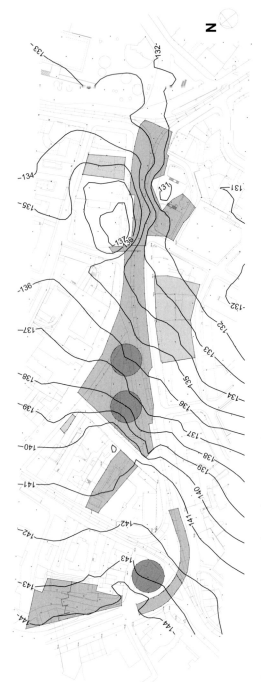

Fig. 8.17. SNP square, contour lines of groundwater level in the 4th state

elevation in groundwater level after the 4th stage can be clearly seen on the contours of change in groundwater level (see Fig. 8.18). The contours are plotted with a 0.5 m step. The maximal elevation of the groundwater level is 4 m and it reaches even to the sides. On the NE side of the model there is a 0.5 m elevation. In the wake of any building in the 4th stage there is a decrease in the groundwater level of 1 m. The elevation in the 4th stage is significant enough to require remediation.

8.3 Remediation Models

This is a large group of models which serves to assess the effectiveness of different variants of remediation and to recommend the best option.

8.3.1 Example of Chotebor, Czech Republic

Numerical modelling was the background for the risk assessment analysis which provides a complex assessment of risk factors and it aimed to set a new target of remediation and length. To solve the existing ecological problem we carried out the modelling for an initial state and for a current state of remediation. The planar model of a steady groundwater flow was done by means of the BEFLOW software system (Drahos and Kovarik 1998b). This system uses BEM to solve the partial differential equation of the groundwater flow (see Chap. 5). The transport of pollution dissolved in groundwater was solved by the random walk particle method combined with BEM.

On modelling the transport of pollution we had two alternatives: transport with or without adsorption. The adsorption is ineffective at places where the rock skeleton is fully contaminated, whereas it is still effective at places distant from the pollution source.

First, we set up a hydraulic model of the domain and then we verified it using the known initial state of groundwater flow. This was followed by modelling the transport of pollution dissolved in groundwater for the initial state. The last stage was the modelling of pollution transport for the current state of remediation.

8.3.1.1 *Groundwater Flow Modelling*

The area for the groundwater flow model was chosen so that it included the area from the pollution source to the local groundwater divide along with the area in the direction of the groundwater flow to the base level of erosion which is now the Brevnicky stream. The east and west boundaries of the area were established by means of the documentation of the contours of potential that were in the project.

154   8 Examples of the Use of Models in Practice

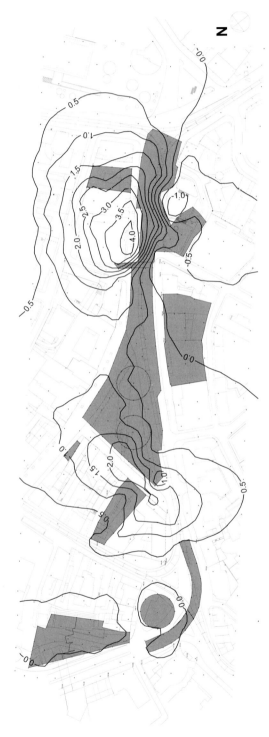

Fig. 8.18. SNP square, contour lines of differences between the 4th and initial states

A schema of the net is in Fig. 8.19. The net is displayed with a simplified topographical map and with homogeneous zones to allow better orientation, and consists of 252 boundary elements.

The boundary elements separate the area of the model from water-bearing migmatic paragneisses. The interzone elements separate the homogenous zones from each other. There are 19 such zones in total. One zone has only one coefficient of hydraulic conductivity and only one thickness. We used only six different coefficients of hydraulic conductivity k and one thickness $M = 16$ m. The values of coefficients of hydraulic conductivity k vary from 1.8×10^{-7} to 59.2×10^{-7} m s^{-1}. The coefficients were set from a previous remediation survey of the locality, and served as values given to the model. During the verification we

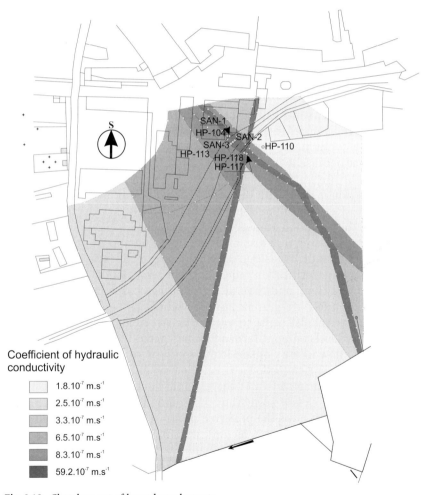

Fig. 8.19. Chotebor, net of boundary elements

adjusted the coefficients k in some of the zones to make the groundwater levels fit the measured values as well as possible. The highest permeability, $59.2 \times 10^{-7}\,\mathrm{m\,s^{-1}}$ was set in the SAN-2 borehole. This value was also assigned to narrow tectonic zones from the geophysical survey. The tectonics was confirmed by a hydrogeochemical survey that indicated that the pollution was spreading from the borehole SAN 2 to the HP 118 which lies approximately at the cross-roads of two tectonic lines (see tectonic lines in Fig. 8.19). The shape of these lines is ruled by the geophysical survey. These lines are the preferred paths of the pollution transport. The tectonic line in the polluted area is bounded by zones with $k = 8.3 \times 10^{-7}\,\mathrm{m\,s^{-1}}$ (see Fig. 8.19). This is an average value from the results of a groundwater pumping from remediation boreholes SAN-1, SAN-3, HP-104 and HP-113. The value of the coefficient k at the place of the borehole SAN-4 is $k = 8.3 \times 10^{-7}\,\mathrm{m\,s^{-1}}$. This value was used in zones Z18 and Z19 (see Fig. 8.19). The value of the coefficient was changed, since we reached a good match between the model data and the measured data in the verification process that used $k = 6.5 \times 10^{-7}\,\mathrm{m\,s^{-1}}$. We consider the first, lower value as strictly local and the second one as characteristic for most of the zone. The bordering zones Z1, Z15, Z17 have only minor tectonic faults and are composed of paragneiss and therefore have small values of the coefficient of hydraulic conductivity $k = 1.8 \times 10^{-7}$ or $2.5 \times 10^{-7}\,\mathrm{m\,s^{-1}}$. Although we used all the surveys and a verification process to set the zone's properties, we cannot expect total precision, which is impossible to reach in a water-bearing fissured medium.

A boundary condition of constant pressure (or of a constant groundwater level in this case because the area is part of a larger area) is given in the boundary elements.

The verification is done with the initial state of the groundwater flow as it is characteristic for most of the whole domain where we have to determine the properties of the pollution transport. The remediation pumping creates an artificial groundwater level in a small area of the domain. We reached a fairly good match between the measured and the determined levels.

The results of the verification model without remediation are the altitudes of the groundwater level and the underground discharge in every boundary element. The altitudes are graphically represented by equipotentials in Fig. 8.20. This picture includes some characteristic streamlines, that together with the groundwater level contours, form a flow net. According to the net, a general direction of the flow in the domain is the south or southsoutheast direction. The more permeable tectonic lines direct the flow only in a narrow band of a faulty zone. The drainage effect of the tectonic lines and the change in flow direction cannot be clearly seen from the contours of potential. The net of streamlines evidently shows the preferred paths of the groundwater flow. The pollution does not spread only in tectonic lines but it can contaminate a wider area. Due to the low permeability of the aquifer, the discharge in the domain is relatively small. The total value is only $0.1614\,\mathrm{l\,s^{-1}}$. The pumping results in an artificial underground discharge, so that the total yield of the remediation wells can be $0.222\,\mathrm{l\,s^{-1}}$.

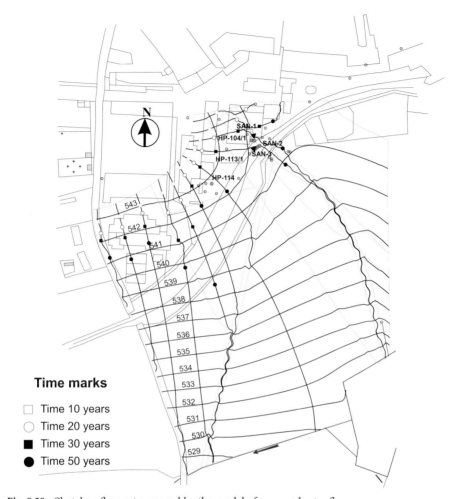

Fig. 8.20. Chotebor, flow net prepared by the model of a groundwater flow

The underground flow discharge is concentrated in the tectonic lines. Thus, a tectonic line emptying into a pond in the southeast part of the domain carries 39% of the total discharge. A tectonic line which empties into a stream in the south of the domain carries another 20% of the total discharge. This means two narrow tectonic lines are involved in 59% of the total discharge from the domain.

To determine the real porous velocity we used the coefficient of effective porosity being 0.13 (according to a background survey).

The first variant to determine the pollution transport is a model of a transport without remediation. The transport of a dissolved pollutant is solved for the initial state. To document effects of the remediation on the pollutant transport, we modelled the transport without adsorption and with linear adsorption.

8.3.1.2
Model of Pollutant Transport Without Adsorption

The main pollution source is placed in the neighbourhood of remediation wells in the southeast corner of the industrial zone. The source was assumed to be permanent. The modelling of the transport of pollution was done for a time period of 50 years. Within this time the front of the pollution reaches the water recipient and 50 years is the time of functioning of the factory and the time of a permanent polluting of the groundwater.

The pollution concentrations determined by the model are given in Fig. 8.21. In 50 years the pollution front reached the pond in an average concentration of 1% (it is a weighted average on the entire width of the breakthrough front).

The average rate of concentration is the average from the displayed width of the pollution's front (approximately 57 m) that reached the pond (Fig. 8.21).

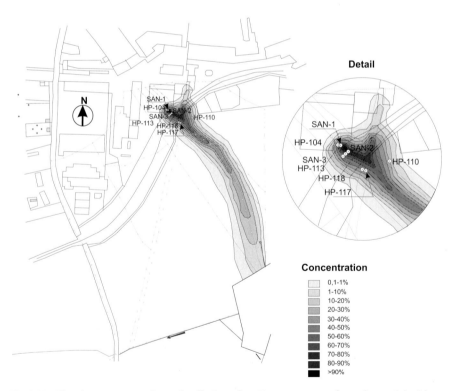

Fig. 8.21. Chotebor, concentrations of pollution after 50 years, output from the model without adsorption

A concentrated narrow band of the tectonic line (7.5 m) causes a recharge to the band of $0.0628 ls^{-1}$ with a concentration rate of 2.5%. The breakthrough curve of an average rate of pollution concentration of the pollution front (57 m) that enters the pond is given in Fig. 8.22. The curve indicates that the pollution might already have reached the pond in low concentrations within 15 years.

Figure 8.21 shows that the pollution dissolved in groundwater is transported along the tectonic line. The width of the main stream of contaminated water with a rate of pollution over 10% is approximately 25 m and so it is wider than the tectonic line which is only 20 m wide. In 50 years a high concentration of pollutants (from 70 to 80%) reaches only the intersection area of the two tectonic lines. This is a relatively small distance from the pollution source. The pollution cloud begins to split directly at the cross-roads of the tectonic lines and the majority follows the tectonic line towards the pond. The smaller part of the cloud sets off to the north to an area of a saddle point of equipotentials. It seems that it flows against the groundwater current. However, after a detailed look one finds out that it is the direction of groundwater discharge into a neighbouring river basin with sources of drinking water. The front of pollution with a concentration of 0.1% spreading to the north moves 75 m from the intersection of the tectonic lines in 50 years.

Here, as well as in other transport models, we presumed the longitudinal dispersion to be 2 m and the transversal dispersion to be 0.2 m in the entire domain.

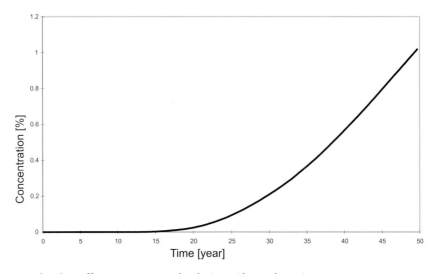

Fig. 8.22. Chotebor, effluent curve to pond, solution without adsorption

8.3.1.3
Model of Transport of Pollution with Adsorption

Chlorinated hydrocarbons dissolved in groundwater have the ability to adsorb on the rock skeleton if its adsorption abilities are not weakened by long-term pollution. The coefficient of distribution stands for a proportion between the concentration of sorbed pollution ($mg\,kg^{-1}$) and the concentration of a dissolved pollution ($mg\,l^{-1}$). Its dimension is ($l\,kg^{-1}$). The retardation factor 3.63 was set with the use of $K_d = 0.125\,dm^3\,kg^{-1}$, $\rho_s = 2.74\,kg\,dm^{-3}$ and $n_{ef} = 0.13$.

The results of the model confirmed the effects of adsorption on slowing down the transport, on decreasing the concentration of the pollution and decreasing the pollution spread. The concentration of dissolved pollution after 50 years of transport can be found in Fig. 8.23. This figure indicates that the concentration of 0.1% at the front of the pollution cloud does not hit the pond. The model proves that within 50 years, pollution with concentrations from 50

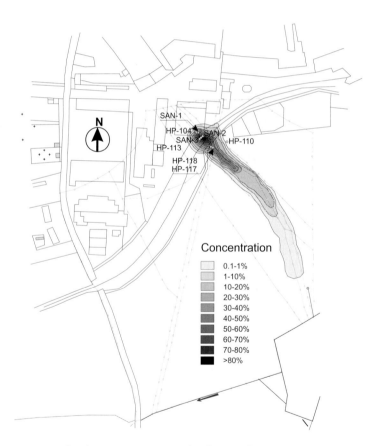

Fig. 8.23. Chotebor, concentrations of pollution after 50 years, output from the model with adsorption

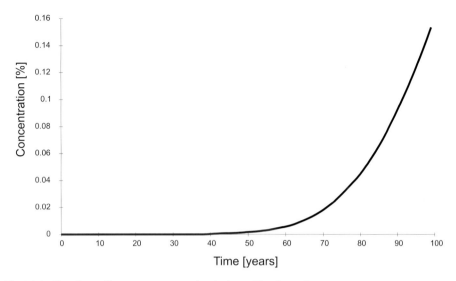

Fig. 8.24. Chotebor, effluent curve to pond, solution with adsorption

to 60% will reach only the intersection of the tectonic lines. This is approximately half the distance of the alternative without adsorption.

The pollution which spreads towards the pond follows the tectonic line and its permeable neighbourhood. The width of the pollution cloud with a concentration of 0.1% is about 35 m (without adsorption it is 67 m). The effects of adsorption can be best observed by comparing Fig. 8.21 with Fig. 8.23. The breakthrough curve in Fig. 8.24 shows that pollution with a low average concentration of 0.0005% can reach the pond in more than 40 years. This curve illustrates an increase in the average concentration after that time. According to this curve the average concentration of pollution in groundwater after 100 years would be 0.16%.

8.3.1.4
Remediation Model

Remediation is carried out by pumping the polluted groundwater by means of wells SAN-1 to SAN-3, HP-104, HP-113 and HP-114. The total yield of all six wells was set to $0.2221 s^{-1}$. The model presumed only one significant pollution source with its position being a machining shop. The model stemmed from the pollution state after 50 years of pollution transport (the model above). Remediation should run for 6 years. The model determining the degree and the spread of pollution was actually done for 56 years (50 years of transport of pollution without remediation and 6 years of remediation). Here we also had two alternatives: with or without adsorption.

8.3.1.5
Remediation Without Adsorption

The concentrations of pollution that express the degree and spread of pollution are shown in Fig. 8.25. The degree and spread of pollution change after 6 years of remediation pumping. The changes can be well observed in Figs. 8.21 and 8.25. A comparison of these pictures indicates that the pollution cloud, in the case of remediation, is torn apart. The tearing-apart is a hydraulic effect caused by the remediation pumping holding the pollution cloud near the intersection of the tectonic lines. The front of the initial cloud was not affected by the remediation actions and it continued to follow the tectonic line towards the pond. This torn-away cloud has a concentration of 30 to 40% in its centre, and the front of the cloud, with a concentration of 10%, moved 45 m towards the pond in the 6 years of remediation, the velocity of the front's move being about 7.5 m per year.

Remediation has a positive influence on slowing down the spread of secondary pollution to the north. Due to the remediation, the front of this cloud was drawn back by 10 m towards the tectonic intersection. The concentration

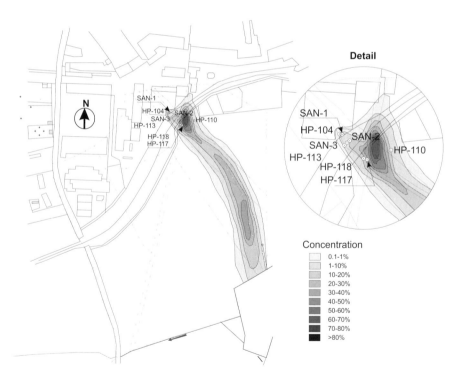

Fig. 8.25. Chotebor, concentration of pollution after 6 years of remediation, output from the model without adsorption

close to the remediation wells decreased to a maximum of 10% in the 6 years of remediation. The centre of pollution is predicted to be between wells HP-117 and HP-110.

8.3.1.6
Remediation with Adsorption

The result of this model (pollution concentrations) is given in Fig. 8.26. The 6 years of remediation with adsorption do not change the spread of pollution so radically, as can be seen in Figs. 8.23 and 8.26. The pollution cloud is again torn

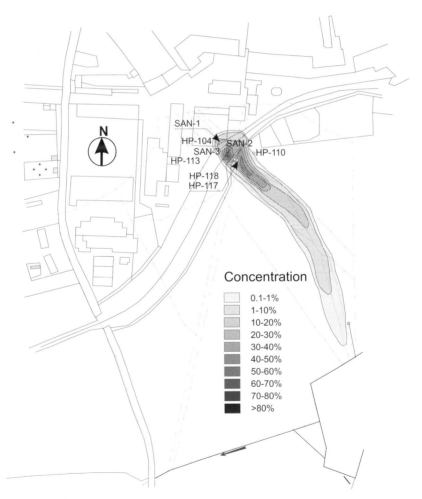

Fig. 8.26. Chotebor, concentration of pollution after 6 years of remediation, output from the model with adsorption

apart at the tectonic intersection. One small centre of pollution is assumed between the borehole HP-113 and a tectonic line of the southsouthwest direction. The centre of pollution has a concentration of 60–70% and the second centre of pollution behind the tectonic intersection has a maximal concentration of 50%. The remediation pumping draws back the secondary cloud's front to southsouthwest.

8.3.2
Example of Airport Bratislava, Slovakia

In this case numerical modelling provides the background that enables the company to continue with the remediation works that eliminate the transport of pollutants in the airport area and at the same time they decrease the pollution in its source.

Due to hydrogeological properties and to an incomplete underground sealing wall, a 3-D model is necessary. The 3-D model of a groundwater flow uses BEM and a model of the transport of pollution uses the DRM method (Drahos and Kovarik 1999).

The surface of a remediation area is a concrete runway of the airport. The groundwater's free surface is about 3.9 m below the surface and an impermeable subsoil is 10 m below the surface at an altitude of 121 m above sea level. The thickness of an aquifer is therefore 7 m. The most contaminated water-bearing rock medium lies between 127 and 128 m above sea level. Even the unsaturated zone above the groundwater level is contaminated to the height that was reached by the groundwater level under extreme conditions. The area with contaminated soil was surrounded by an underground sealing wall. Its low permeability is characterised by its coefficient of hydraulic conductivity $k = 5 \times 10^{-8}\,\mathrm{m\,s^{-1}}$. The top edge of the wall has an altitude of 130.3 m above sea level and the bottom edge has an altitude of 124.3 m above sea level. The sealing wall is hydraulically incomplete as the bottom border of the aquifer is at 121 m above sea level. The ground plan of the wall has the form of an irregular hexagon (see Fig. 8.27). The former underground fuel tanks can be seen there which were a primary source of the pollution of the rock medium. The tanks have already been removed and the main source is now the contaminated soil. Figure 8.27 depicts the position of an extracting remediation well HSJ-1 and positions of observation wells HM-1 to HM-5. An outer observation borehole, HM-4, is situated to the southsoutheast of the centre of pollution behind the outer edge of the sealing wall. There is one more remediation borehole, HLS-3a, behind the southwest edge of the wall, but it is not relevant in this model. The aquifer's coefficient of hydraulic conductivity was set by surveys to $k = 2.1 \times 10^{-3}\,\mathrm{m\,s^{-1}}$. The vertical permeability was set at one order lower than the value above. The effective porosity is 0.2 and the coefficients of longitudinal and transversal dispersivity are 1 and 0.1 m, respectively.

The method's net stretches sufficiently outside the polluted area and can be observed in Fig. 8.28. The 3-D net consists of three horizontal layers of aquifer.

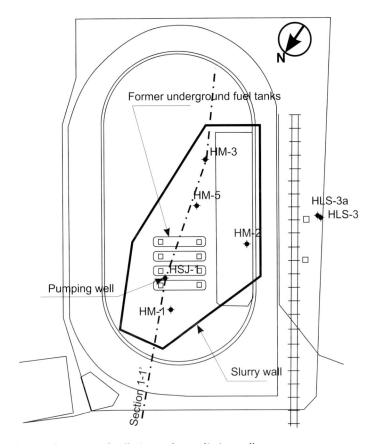

Fig. 8.27. Bratislava, layout of a source of pollution and remediation wells

The top layer represents a polluted 1-m-thick layer of water-bearing gravel. The top boundary of this layer is the free surface of the groundwater. The middle layer has its bottom boundary at the same altitude as the bottom edge of the sealing wall. The middle layer was designed to match the hydraulic incompleteness of the sealing wall. The bottom layer is based in an impermeable subsoil. There is a boundary condition of the 3rd kind given on the sides of the net that simulates the presence of a surrounding water-bearing medium.

8.3.2.1
Initial State of Groundwater Flow with Sealing Wall

The 3-D model was set up to determine the potential values (piesometric levels) and the level of a free surface with a sealing wall but without pumping. The determined piesometric levels at the bottom of the aquifer are depicted by

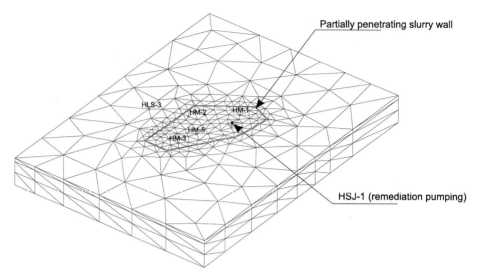

Fig. 8.28. Bratislava, spatial view of the net of elements

means of contours of potential in Fig. 8.29. It seems, from the picture, that the flow of the groundwater is uniform at the bottom of the aquifer, and the field of velocity is therefore homogenous. Figure 8.30 shows, using the contours of potential, the properties of the groundwater flow at the free surface level. A barrier effect of the sealing wall at this level has already showed itself by curving the equipotentials and by changing the character of the field of velocity.

8.3.2.2
Pumping Only from HSJ-1

Pumping from the borehole HSJ-1 $18 \mathrm{l s}^{-1}$ was the action that took place prior to any modelling. Thus the model had to solve this variant of remediation first. The groundwater flow is again represented by contours of potential which can be seen for an altitude of 121 m above sea level in Fig. 8.31 (i.e. the bottom of the aquifer). Figure 8.32 is the view of contours at the free surface level. The pumping device is positioned over the bottom of the borehole at 124.4 m above sea level, where the groundwater pumping is concentrated. A comparison of Figs. 8.31 and 8.32 shows that the pumping and the draining effects of the HSJ-1 borehole are greatest at the groundwater level, due to the sealing wall. The comparison above also shows that the potential at the bottom of the aquifer is higher than the one at the free surface level. This fact causes an upward flow of groundwater. The upward flow of relatively clear water through a contaminated area has a positive effect on clearing the water-bearing rock medium. This effect is caused by building the underground sealing wall, which was constructed as an incomplete barrier element.

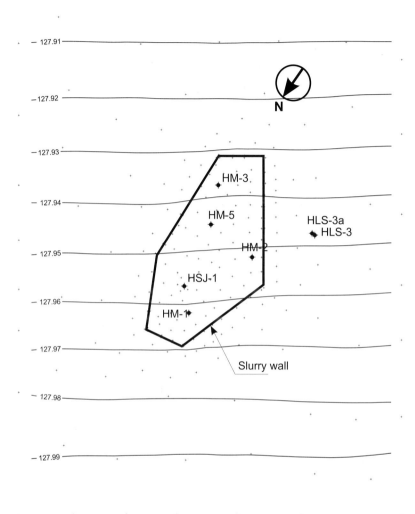

Fig. 8.29. Bratislava, equipotentials in a bottom layer in an initial state

8.3.2.3
Common Pumping from HSJ-1 and HLS-3a and Reinjecting of Treated Water into HM-2 and HM-3

One possible remediation method is a reinjection of the treated water back into the rock medium. For this reason the other variant of the groundwater flow model includes a simultaneous pumping from extracting wells and the reinjection of half of the pumped water back into the aquifer. The equipoten-

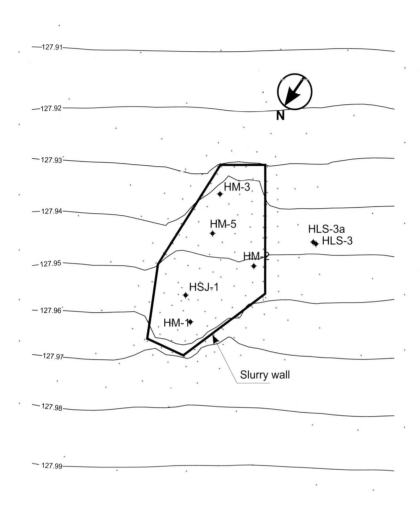

Fig. 8.30. Bratislava, contours of the groundwater level in an initial state

tials of simultaneous pumping and recharging are given in Fig. 8.33. By recharging the cleaned water into HM-2, we create a groundwater divide between wells HSJ-1 and HLS-3a that does not allow free movement of pollution from one borehole to the other. Pollution from the neighbourhood of HLS-3a cannot reach HSJ-1 when pumping large amounts of water from HSJ-1.

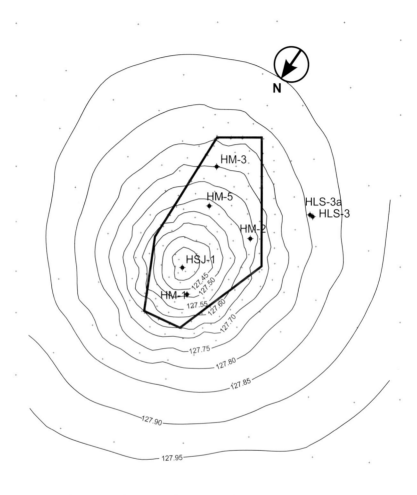

Fig. 8.31. Bratislava, equipotentials in a bottom layer during a remediation pumping

8.3.2.4
Transport of Pollution in the Initial State of Flow and in the State with Sealing Wall

A longitudinal section 1 – 1′ of the state of groundwater pollution after 1 year from building of the sealing wall is given in Fig. 8.34. According to this picture,

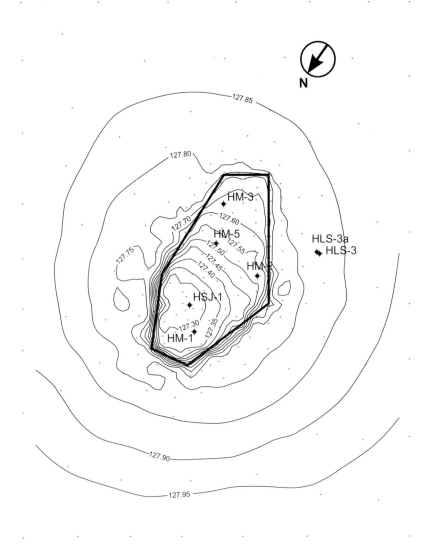

Fig. 8.32. Bratislava, contours of the groundwater level during a remediation pumping

pollution does not get behind the wall in 1 year's time. A planar flux of the pollutant's mass is presumed to be orthogonal to the groundwater surface and results in a maximal concentration of $3\,\text{mg}\,\text{l}^{-1}$ of oils on the surface. The intensity of the mass flow decreases with time. The decease is regular and the intensity reaches one half its initial value in 5 years. Figure 8.34 shows that the pollution is concentrated in the part of the sealing wall that intersects the groundwater flow.

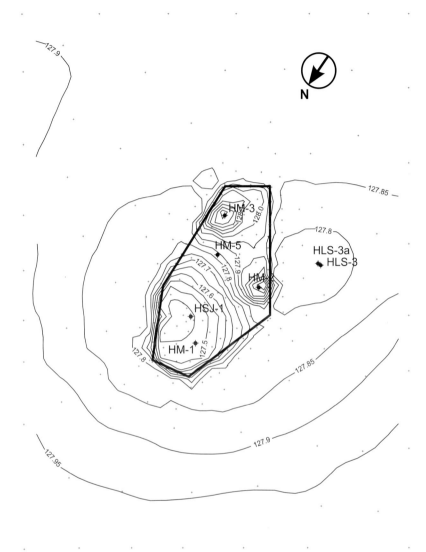

Fig. 8.33. Bratislava, contours of the groundwater level during pumping and injection

Figure 8.35 shows the same section $1 - 1'$ as above but for a time period of 5 years. After this time the pollution will have spread behind the sealing wall. There is a concentration of $2.5\,\mathrm{mg\,l^{-1}}$ of the pollutant in the groundwater behind the wall in the direction of flow.

The solution implies that the pollution sticks to the area surrounded by the sealing wall only for a year or so. After 1 year without remediation the pollu-

172   8 Examples of the Use of Models in Practice

Section 1-1

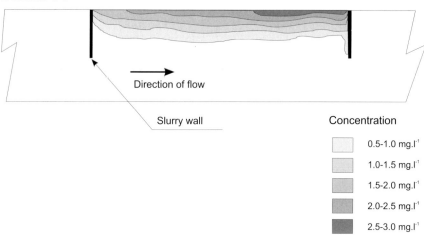

Fig. 8.34. Bratislava, contours of the pollution in the initial state after 1 year

Section 1-1'

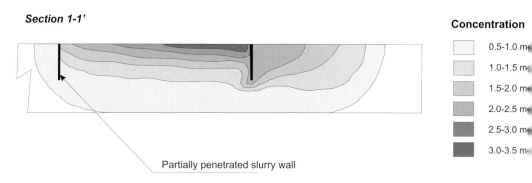

Fig. 8.35. Bratislava, contours of the pollution in the initial state after 1 year

tion leaves the area in the direction of the groundwater flow. Remediation is therefore necessary.

8.3.2.5
Transport of Pollution with Pumping from HSJ-1

We begin from a pollution state after 1 year since completion of the sealing wall, then we add a simulation of pumping from the well HSJ-1 lasting 2 months. After a longer time the pollution does not change, which means a longer remediation has no great effect on the pollution's concentration. This quasistable pollution is in the 1 – 1' section in Fig. 8.36. The comparison tells us that after 2 months a relatively large yield of $18 \, l \, s^{-1}$ causes a reduction in the pollutant's concentration by half. Continuing the remediation leads only to

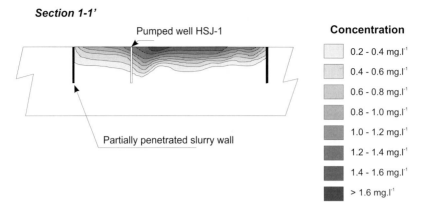

Fig. 8.36. Bratislava, contours of the pollution after the remediation pumping

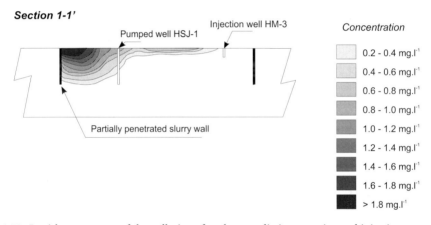

Fig. 8.37. Bratislava, contours of the pollution after the remediation pumping and injection

a marginal decrease in concentration, and so long-term pumping from the HSJ-1 well is ineffective. On the other hand, stopping the pumping can lead to a spread of pollution out of the area as in Section 8.3.2.4.

8.3.2.6
Transport of Pollution with Pumping from HSJ-1 and HLS-3a and Recharging into HM-2 and HM-3

We again began with the same state as in the previous paragraph. The quasi-stable concentration value is reached already after 21 days of a simultaneous pumping and recharging. The spread and the degree of pollution can be found in Fig. 8.37, which documents a fairly irregular spread of pollution. The aquifer surrounded by HSJ-1 and HM-3 and HM-2 is cleared up to a concentration level of $1\,\mathrm{mg\,l^{-1}}$. On the opposite side of the HSJ-1 well, there is a concentration maximum of over $1.6\,\mathrm{mg\,l^{-1}}$. The pollution after 21 days is on average lower than after pumping only from the HSJ-1 well.

All examples listed in this chapter illustrate the use of numerical methods in hydrogeology. We focused on the BEM and FEM methods because the finite differences method is widely used and well known, unfortunately also in cases where it is either not suitable or ineffective.

References

Bansky V, Kovarik K (1995) Vybudovanie hydraulickej clony pre ochranu zdroja pitnej vody, štúdia. Manuscript, archive Geosoft, Žilina

Bondarenkova Z (1993) Žiar n/H, ochranné pásmo vodného zdroja. Final report, archive Geofond Bratislava

Drahos M, Kovarik K (1998a) Modelovanie prudenia podzemnej vody na ulohe Dostavba namestia SNP. Final report No. 3/98, archive FMG, Banska Bystrica

Drahos M, Kovarik K (1998b) Modelovanie sanacie na lokalite Chotebor. Final report No. 4/98, archive FMG, Banska Bystrica

Drahos M, Kovarik K (1999) Letisko Bratislava, matematicky model sanacie. Final report No. 2/99, archive FMG, Banska Bystrica

Kovarik K (1994) Priestorový model prúdenia podzemnej vody v okolí VD Žilina. Manuscript, archive Geosoft, Žilina

9
Description of Software Included in This Book

The CD included in this book contains simple versions of software using some of the methods presented in Chapters 5 and 6. These academic versions serve to demonstrate the algorithms of the method. The algorithms themselves are shown in a source code for a better understanding of the principles, and this chapter was written to make comments on the algorithms. As we have already mentioned several times, a program dealing with the preparation of input data and with a visualisation of the results is an important part of every software system and such a program was added to every system mentioned.

All software was developed by means of C++ programming language namely by Visual C++ version 6.0 from Microsoft Corp. and it works on Windows 95, 98 and NT4.0 operation systems (it uses a Win32 library only for 32-bit operation systems).

9.1
BEFLOW System

This is a system using the boundary element method for planar models of a steady groundwater flow (see Chap. 5.1.4) and the random walk method (see Chap. 6) for models of pollution transport. The system consists of five programs and every one of them can be run independently. These programs share data using a disk file. To work with the disk the whole system uses serialisation, which means that all classes are derived from the CObject class and are stored on the disk or read from the disk with the help of the Serialize subprogram (see e.g. Kruglinski 1998).

The system's most important classes are as follows

- *CNode* – class of nodes. Elements of this class are not nodes themselves (as in the case of constant base functions the centre of an element) but the beginning and the end of an element.
- *CElement* – class of elements. This class stores numbers of the beginning and the end of an element, the type of an element, the value of a boundary condition as well as the value of a potential and its derivative.
- *CIntPoint* – class of inner points. The program determines the value of a potential and values of the velocity in the direction of each axis in every inner point. The data along with coordinates of the point are stored for every inner point.

Definitions of these classes in C++ are given in Table 9.1. In addition to what we have mentioned above, each class contains functions that use elements of this class such as, for example, functions that plot the class's element or compare and transform the coordinates. Table 9.1 lists only a minimum of what the classes have to contain. This is the data needed for a solution by the program BemSolve. Other plotting programs add the drawing of the class's

Table 9.1. Declaration of classes of the system BEFLOW

```
// Class of nodes
class CNode:public CObject
{
public:
        CNode (float x,float y);
protected:
        CNode();
        DECLARE_SERIAL (CNode);
public:
        void Coord (float& x,float& y);
        float xb,yb;           // xb, yb are the coordinates of node
        virtual void Serialize (CArchive& ar);
};
// Class of elements
class CElement:public CObject
{
public:
        CElement (int u1,int u2);
protected:
        CElement();
        DECLARE_SERIAL (CElement);
public:
   int nod1,nod2,type;   // nod1, nod2 are number of nodes, type is the type of boundary
   condition
     float pot,der;        // pot is a value of potential, der is a value of its derivative
     float level,kf0,b0;   // level, kf0, b0 are the data about the b.c. of the 3rd kind.
     virtual void Serialize (CArchive& ar);
};
// Class of wells
class CWell:public CObject
{
public:
   CWell (CString s,float x,float y,float *gc,float dis,float d);
protected:
        CWell ();
        DECLARE_SERIAL (CWell);
public:
        float xz,yz,yield,diam,drwdown;
        CString name;
        virtual void Serialize (CArchive& ar);
};
```

elements and the coordinates' transformation to this definition. Programs that depict the results in a graphic form on the screen use a basic class CZoomView (see Pirtle, 1995 or www.codeguru.com). This class also enables zooming of a selected part of the program's window.

9.1.1
Program BemInp

This program is aimed to prepare the input data. It can be found on the CD only in its EXE form. This section is therefore meant to be a guide for the proper use of this program.

The program's window (Fig. 9.1) is composed of a main menu and a tool bar with icons below. There is a status bar at the bottom of the window where short memos are written.

Other windows for different documents can be opened in the main working area which means that we can open more documents at the same time. One document stands for one set of input data. The program is directed by means of commands either from the main menu or from pulldown menus. Some commands that are more frequently used have an icon associated with them to speed up the work.

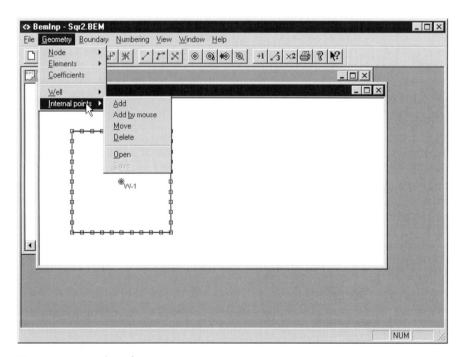

Fig. 9.1. Main window of program BemInp

Table 9.2. Commands of the main menu of program BemInp

Command	Function
File	Opens a pulldown menu that allows you to work with disk files that contain input data. It also contains commands that print the net of elements
Geometry	Opens a menu for input of the geometric shape of the area
Boundary	Opens a menu for input of the boundary conditions
Numbering	Commands from this menu activate or deactivate the numbering of nodes
View	This menu is used for showing or hiding the tool and status bar and for zooming
Window	These commands enable you to arrange multiple views of multiple documents in the BemInp window
Help	This command provides help with this application

Table 9.3. Commands of the *File* menu of program BemInp

Command	Function
New	Creates a new blank window where you can prepare a new task
Open	Opens an existing document that is stored in the disk file
Save	Saves data about a created net of elements to the same disk file
Save as	Saves data about a created net of elements to a specified disk file
Print	Prints the net of elements on the printer
Print preview	Displays the net of elements on the screen as it would appear printed
Print setup	Selects a printer and a printer connection
Exit	Exits BEMINP

9.1.1.1
Commands of Program BemInp

This paragraph is intended as a short review in form of a table (Table 9.2) that should serve for a quick orientation in the program's commands. If there is no opened window for data preparation, the main menu is reduced only to commands *File*, *View* and *Help*. To obtain a complete menu you have to either open an already existing document or create a blank window with the help of the *New* command. After the regular start of the program, the new window is created automatically.

9.1.1.2
Commands of the File Menu

These commands serve to work with disk files containing input data as well as for printing of a net created from the input data. All the commands are listed in Table 9.3. They belong to standard commands used by almost all programs running in the Microsoft Windows environment. Thus we mention them only briefly.

As we have stated before a new window in the main window of the program is used for input of the data. This window is created by the *New* command.

The command *Open* is then referred to when we have an already existing set of data that is stored in a disk file. First, a standard dialog box is opened to choose a file you intend to work with. The files are created by means of serialisation and so they are binary. They can be created and read only by the BEFLOW system. The files have an extension BEM. After you pick a file, the command *Open* opens it, it reads the data and then it creates a window to display the net of elements along with inner points and sources that it plots using the data from the file. The name of a file is written in the heading of a window.

The command *Save* saves data from a currently active window into a disk file. If this window is created by the command *Open*, all new data will be saved in the file which we opened; but if the window is created by the *New* command, you have to type the name of the file where the data should be saved (again with the help of a dialog box). If you use *Save as* instead of Save, you have to give the name in both cases.

When you choose the *Print* command, it opens a dialog box to set up the properties of print. Here you can choose the type of printer used. The program BemInp prints always only one page with the net of elements, hence there is no possibility of choosing the pages which should be printed.

To change the type of printer, the *Print Setup* command can be used and additionally, this command allows you to change the page format and the page's orientation. All the settings done by these commands are only temporary and after exiting the program they return to the initial settings. To make these changed settings permanent, you have to change them directly in the Windows system.

The command *Exit* finishes the program and closes all windows. The command tests whether you have saved all the changes into a file and if there are any not-saved changes, the program asks you whether they should be saved before exiting.

9.1.1.3
Commands of Geometry Menu

An important set of commands is that which creates the net of elements. When you click on the *Geometry* written in the main menu, a pop-up menu appears with commands which are in Table 9.4. All commands that use a mouse can be deactivated by clicking the right button of the mouse.

Here are the commands in an order in which they are used by a setup of a new net. First, one has to set the coordinates of nodes. Obviously, we use the *Nodes* command which results in a new pulldown menu with commands which are in Table 9.5.

Table 9.4. Commands of the *Geometry* menu of program BemInp

Command	Function
Nodes	A group of commands for setting the nodes which are the points at the end of an element
Elements	A group of commands for setting the elements
Coefficients	Here you can set the coefficients of hydraulic conductivity and the thickness of aquifer
Int.points	A group of commands to set the inner points inside the domain
Wells	A group of commands to set the wells

Table 9.5. Commands of *Nodes* menu

Command	Function
Add	Adds a node by giving its coordinates
Add by mouse	Adds a node using a mouse
Move	Moves a node into a new position
Delete	Deletes a chosen node
Open	Opens a text file containing coordinates of nodes
Save	Saves the coordinates of nodes in a text file

9.1.1.3.1
Command Nodes

The command *Add* opens a simple dialog box where you can type the coordinates x, y of a new node. This dialog opens again and again and thereby you can add further nodes. To finish the addition of nodes just click on the *Stop* button (or press *Esc* button on the keyboard).

Nodes can be added also by the next command *Add by Mouse*. In this case a node is added in the position of the mouse's cursor by clicking the left button. This function can be used only after three nodes have already been given because only then can the computer transform the real coordinates into the coordinates of the screen. While using this function, there are real coordinates of the mouse cursor shown on the left side of the status bar.

The position of an already given node can be altered by means of the *Move* command. This again requires the use of a mouse. Place the cursor on the node whose position you want to change and click the left button and holding a pushed button drag the mouse across the screen. There are again real coordinates of the cursor on the left side of the status bar. Releasing the button fixes the node's new position.

To remove a node one must use the *Delete* command. Just place the cursor on the node and click the left button. The program first asks for your confirmation and then it removes the node.

The commands *Open* and *Save* work with text files where there is a couple of numbers on each line which stands for the node's coordinates. The *Open* command first opens a dialog box which enables to pick a file with the coordinates and then it adds the coordinates to those used before opening the file. The *Save* command also opens a dialog box to give the name of a file where the program stores the coordinates of all nodes.

9.1.1.3.2
Command Elements

The command *Elements* from the *Geometry* menu also results in another menu containing commands shown in Table 9.6.

The *Add* command serves to add only one element: one has to give the nodes at the beginning and at the end of an element by clicking the left mouse button with the cursor on those two points. Afterwards, an element is created and drawn on the screen. A more suitable option to add elements is to use the second command *Add All* where one has to give the node at the beginning and the node at the end of an element for the first element and only the node at the end for the rest of the elements (the node at the beginning being assumed identical with the node at the end of the previous element). To finish this activity just click the right button.

To remove an element use the *Delete* command. After activating this function place the cursor on an element and push the left button. The program asks for your confirmation and thereafter it removes the element.

One has to be cautious when adding the elements. The best way to add elements is to use the *Add All* command which adds them as a complete series and to add them on the outer boundary in an anti-clockwise sense and on the inner boundary (a hole in the domain) in a clockwise sense. This way you can be sure to have a correct polarity of the normal derivative. The order does not have to be kept when correcting a single element, whereas one has to keep the orientation of the element. Not fulfilling this condition can lead to incorrect results. As the orientation of elements cannot be checked easily, the program does not attempt to do so.

Table 9.6. Commands of *Elements* menu

Command	Function
Add	Adds one element
Add all	Adds a group of elements
Delete	Removes an element

Table 9.7. Commands of *Well* menu

Command	Function
Add	Adds a well by giving its coordinates
Add by mouse	Adds a well using a mouse
Edit	Edit data of a chosen well
Move	Moves a chosen well into a new position
Delete	Deletes a chosen well

9.1.1.3.3
Command Coefficients

After having set the elements, it is time to give the coefficients of hydraulic conductivity and the thickness of an aquifer. This is done by selecting the *Coefficients* which opens a dialog box where coefficients' values k_x, k_y should be given along with the thickness of an aquifer. Otherwise all these values are automatically set to 1.

9.1.1.3.4
Command Wells

The command *Wells* serves for an input of point sources or sinks. Every source is given by its name, coordinates, yield and diameter. The name is for identification purposes and the coordinates and the yield are needed for a calculation. The diameter is used for determining the potential's value directly in the source. In BEM the potential in the source is not defined and it converges to an infinite value (see Chap. 5). Thus the potential's value is determined in a distance that equals a half of the diameter from the source's centre. In the case of a borehole it is a potential on the borehole's perimeter. To work with wells, we use a menu with commands listed in Table 9.7.

The command *Add* results in a dialog box (see Fig. 9.2) where you can give the name of a source, its coordinates x, y, yield and its diameter. *Add by mouse* works the same way as the command in the menu *Nodes*. The name is automatically generated in a form W-n where n is the number of the source, the yield is automatically set to zero and the diameter to 0.1. All these data can be changed by *Edit* after placing the cursor of your mouse on the source whose data should be changed and clicking the left button.

There are other commands that work as those in the *Nodes* menu such as *Move* and *Delete*. Hereby you have finished defining the geometric shape of the domain. You can give additional inner points (points that lie inside the domain) and as BEM does not use these points in the solution (they are not a part of the system of equations), you can add them whenever you like and do not have to set up the system of equations and to solve it anew. These points serve only as additional information about the potential and about the velocity of the flow. That is what you have to determine after adding such a point (see Section

Fig. 9.2. Dialog box for input data of well

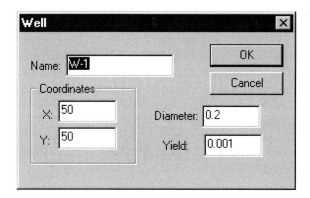

9.2.2). There is a very similar menu to the *Nodes* to put in information about inner points (see Table 9.5).

9.1.1.4
Commands of Boundary Menu

After setting up the net of elements, one has to define the boundary conditions. That is what is done in this menu. One boundary condition must be given for each element. One has to set a boundary condition of the 1st kind at least in one element since the BEFLOW system works with a steady flow.

The *Boundary* command leads to a menu that has only two functions.

New – serves to set a new boundary condition in an element.
Edit – serves to change a boundary condition in a chosen element.

The function *New* setting a new boundary condition can set it on the entire boundary (command *All*) or only in some elements (command *In Element*). After using both of these commands, a dialog box appears (see Fig. 9.3) where you can give all needed values.

Potential – Here one can set the potential's value for a boundary condition of
   the 1st kind or the groundwater level for a boundary condition the 3rd kind
   (see Chap. 5.1.4.8).
Derivative – Here one should set the value of the potential's exterior normal
   derivative for a boundary condition of the 2nd kind.
Distance – Here one can set the distance for a boundary condition of the 3rd
   kind.
Coefficient – Here one can set the coefficient k_0 for a boundary condition of
   the 3rd kind.
Kind of condition – It is quite obvious what is set here. There is a choice of
   three conditions: The Dirichlet (1st kind), Neumann (2nd kind) and Cauchy
   (3rd kind). The values of a boundary condition of the 3rd kind are terms of
   this equation

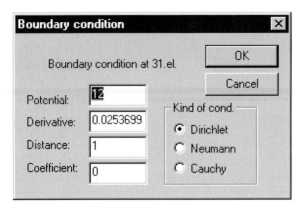

Fig. 9.3. Dialog box for input of boundary conditions

$$q = \frac{k_0}{T_x b_0}(\overline{H} - u). \tag{9.1}$$

This stands for a relationship between the potential and the flux in a given element. This condition can be used to match effects of a clogged river bed and in that case: k_0 is the coefficient of hydraulic conductivity of the clogged layer at the river bottom, b_0 is the thickness of the layer, $\overline{H}$ is the water level in the river.

This condition can also be used to distance a boundary condition from the boundary of a domain. This is used for a solution of a selected part (a detail) of a domain. The condition's values are given either to every element (command *All*) or only to elements selected by a click of the mouse (command *In Element*). The command *Edit* works similarly. After clicking on the element of your choice, a dialog box opens where all values can be altered.

9.1.1.5
Commands of Numbering Menu

This menu activates (or deactivates) the numbering of elements, nodes or inner points. The numbers of nodes are Nxxx, the numbers of elements are Exxx and the inner points' numbers are Pxxx.

9.1.1.6
Commands of View Menu

All the functions of this menu are listed in Table 9.8. These are only secondary functions that effect displaying the results on the screen.

The tool bar is composed of icons that are there as a hot key to the most important functions. Its standard position is below the main menu but generally it can be anywhere in the window. The *Toolbar* command only shows and

Table 9.8. Commands of *View* menu

Command	Function
Toolbar	Shows or hides the toolbar
Status Bar	Shows or hides the status bar
Zoom in	Zooms in the picture on the screen
Zoom out	Zooms out the picture on the screen
Zoom full	Plots the picture in the initial size

Table 9.9. Commands of *Window* menu

Command	Function
New	Creates a new window that views the same document
Cascade	Arranges windows in an overlapped fashion
Tile	Arranges windows in non-overlapping tiles
Arrange icons	Arranges icons of closed windows

hides it and it does not change the position. If the tool bar is shown, there is a tick by the command. The status bar serves for comments such as help to the current menu's options or a display of coordinates of the mouse cursor (by some functions). A default status bar's position is always at the bottom of the window.

The next three functions change the view of the pictures on the screen. The first one (*Zoom in*) changes the cursor to a magnifying glass and activates the zooming-in. Every click of the mouse increases the zoom two times. If there is a particular part of the screen we want to zoom, we place the cursor in the left upper corner and then we drag the mouse with the left button pushed across the screen to the right bottom corner. Afterwards, this section appears in the whole window. This function is deactivated by a click of the right button of the mouse. The *Zoom-out* function does the complete opposite to the *Zoom-in* and each click of the mouse decreases the zoom by half. In order to return to the initial size one should use the *Zoom-full* command.

9.1.1.7
Commands of Window Menu

There is again a table listing the commands of this menu (see Table 9.9). These commands are regularly used in all Windows applications. They allow the user to sort different opened windows in the main window of the program BemInp.

9.1.1.8
Commands of Help Menu

These commands guide an inexperienced user through the program.

9.1.2
Program BemSolve

This is the program that does the actual solution of a planar groundwater flow using BEM and it displays the results and prints the input data.

9.1.2.1
Basic Relations

The program works with the simplest base functions (constant polynomials) in every element which has therefore only one node in its middle. The basic equation of this method has the form of Eq. (5.58) and in a matrix form it is Eq. (5.60). The terms of the matrices **G**, **H** are defined as

$$G_{ij} = \frac{1}{2\pi} \int_{\Gamma_j} \ln\left(\frac{1}{r_{ij}}\right) d\Gamma_j$$

$$\tilde{H}_{ij} = -\frac{1}{2\pi} \int_{\Gamma_j} \frac{D}{r_{ij}^2} d\Gamma_j, \quad (9.2)$$

where r_{ij} is a length of the vector that begins at point i and runs through the whole element Γ_j and D is an orthogonal distance between the point i and a line that is a support of the element Γ_j [see Eq. (5.62) and Fig. 5.13]. To calculate the integrals in Eq. (9.2), the program uses their analytic solution that shows a higher stability at critical places (when determining the potential in inner points) than a numerical solution. In the analytic solution, we use the following expression instead of r_{ij}

$$r_{ij} = \frac{D}{\cos\varphi}. \quad (9.3)$$

The differential $d\Gamma_j$ can be written as (see Fig. 9.4)

$$d\Gamma_j = \frac{r_{ij} d\varphi}{\cos\varphi} = \frac{D}{\cos^2\varphi} d\varphi \quad (9.4)$$

after a substitution into the first integral in Eq. (9.2), we obtain

$$G_{ij} = -\frac{D}{2\pi} \int_{\varphi_1}^{\varphi_2} \left(\frac{\ln D}{\cos^2\varphi} - \frac{\ln\cos\varphi}{\cos^2\varphi}\right) d\varphi. \quad (9.5)$$

This integral can be split into a sum of two integrals. The first one can be solved directly and the second one by means of the by parts method. The result is

$$G_{ij} = -\frac{D}{2\pi}[\tan\varphi_2(\ln r_{i2} - 1) - \tan\varphi_1(\ln r_{i1} - 1) - \varphi_1 + \varphi_2], \quad (9.6)$$

where r_{i1} is a distance of the beginning of the element Γ_j to the point i and r_{i2} is a distance from the end of the element Γ_j to the point i. The second integral in Eq. (9.2) can be solved analogously and we obtain

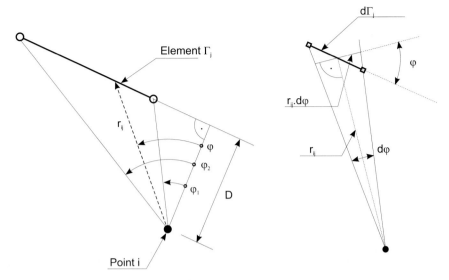

Fig. 9.4. Integration over the boundary element

$$\tilde{H}_{ij} = \frac{1}{2\pi} \int_{\varphi_1}^{\varphi_2} d\varphi. \tag{9.7}$$

After integration we acquire this simple relation

$$\tilde{H}_{ij} = \frac{1}{2\pi}(\varphi_2 - \varphi_1). \tag{9.8}$$

If the point i lies on the line that is a support of the element Γ_j, we obtain this relation after a limit transition

$$G_{ij} = -\frac{1}{2\pi}[r_{i2}(\ln r_{i2} - 1) - r_{i1}(\ln r_{i1} - 1)] \quad \tilde{H}_{ij} = 0. \tag{9.9}$$

The whole integration is done by a subprogram Integ which is a part of a class CBemSolveDoc (see Table 9.10). The parameters of the subprogram are:

x_p, y_p – transformed coordinates of the point i.
x_1, y_1 – transformed coordinates of the beginning of the element Γ_j.
x_2, y_2 – transformed coordinates of the end of the element Γ_j.
h_p, g_p – terms H_{ij} and G_{ij}. These are the output parameters and are produced by the program. Both parameters are referred to by their name and not by the value.
tt – the Jacobi determinant of the coordinates' transformation to fit a medium's anisotropy.

The Jacobi determinant in this case has a simple form.

Table 9.10. Subprogram Integ

```
void CBemSolveDoc::Integ (double xp,double yp,
         double x1,double y1,double x2,double y2,
         double& hp,double& gp,
         double tt)
{
double r1,r2,r,a2,b2,c2,dx1,dx2,dy1,
         dy2,ln1,ln2,xo1,yo1,yo2,yop,xop,nx,ny,
         pi=3.1412569,dist,pi2,fi2,fi1,tg1,tg2;
pi2=0.5*pi;
dx1=x1−xp;dx2=x2−xp;dy1=y1−yp;dy2=y2−yp;
a2=dx1*dx1+dy1*dy1;b2=dx2*dx2+dy2*dy2;
dx1=x1−x2;dy1=y1−y2;
c2=dx1*dx1+dy1*dy1;
r1=sqrt(a2);r2=sqrt(b2);r=sqrt(c2);
ln1=log(r1);ln2=log(r2);
nx=(y2−y1)/r;ny=(x1−x2)/r;
xo1=nx*x1+ny*y1;xop=xp*nx+yp*ny;
dist=xo1−xop;
yo2=−ny*x2+nx*y2;yo1=−ny*x1+nx*y1;yop=−ny*xp+nx*yp;
if (fabs(dist)<1e−8)
{
    r2=yo2−yop;r1=yo1−yop;
    gp=−tt*(r2*(ln2−1)−r1*(ln1−1));
    hp=0;
}
else
{
    tg1=(yo1−yop)/dist;tg2=(yo2−yop)/dist;
    fi2=atan(tg2);fi1=atan(tg1);
    if (fabs(fi2)==pi2||fabs(fi1)==pi2)
    {
        r2=yo2−yop;r1=yo1−yop;
        gp=−tt*(r2*(ln2−1)−r1*(ln1−1));
        hp=0;
    }
    else
    {
        hp=−tt*(fi2−fi1);
        gp=−dist*tt*(tg2*(ln2−1)−tg1*(ln1−1)
             −fi1+fi2);
    }
}
}
```

$$tt = \det\begin{bmatrix} \dfrac{\partial x}{\partial \tilde{x}} & \dfrac{\partial x}{\partial \tilde{y}} \\ \dfrac{\partial y}{\partial \tilde{x}} & \dfrac{\partial y}{\partial \tilde{y}} \end{bmatrix} = \det\begin{bmatrix} 1 & 0 \\ 0 & \sqrt{\dfrac{T_y}{T_x}} \end{bmatrix} = \sqrt{\dfrac{T_y}{T_x}}. \tag{9.10}$$

This subprogram is called to determine the terms and to set up the matrices **H, G** as well as to determine the potential in inner points of the domain.

Table 9.11. Boundary conditions in the program BemSolve

```
for (i=0;i<nel;i++) dl[i]=drs[i]=0.0;
for (i=0;i<nel;i++)
{
    CElement* el=GetElement(i);
    switch (el->type)           // kind of b.c. in the i-th element
    {
    case 1:
        dl[i]=el->pot;
        break;
    case 2:
        dl[i]=el->der;
        break;
    case 3:
        qin=el->kf0/(el->dist*tx);
        dl[i]=qin*el->level;
        drs[i]=qin;
        break;
    }
}
```

The setup of the system of equations itself is done by a subprogram Assemble in the class CBemSolveDoc. First, we have to prepare a vector to determine the right side of the system of equations depending on the kind of the boundary condition (see Table 9.11).

The integration of the matrices **H** and **G** itself is done by a nested loop (see Table 9.12). In C++ programming language the matrices **H** and **G** were declared in the extended memory as vectors thus the program uses a function Ind to determine the index of the term of a matrix.

An algorithm for setting up the matrices, sets up the diagonal terms of the matrix **H** so that they equal a negative sum of non-diagonal terms of one line.

$$H_{ij} = -\sum_{j=1, j \neq i}^{N} H_{ij}. \tag{9.11}$$

Every term of the matrix **H** is increased by a difference that is caused by a possible boundary condition of the 3rd kind (see Chap. 5.1.4.8).

$$\tilde{H}_{ij} = H_{ij} + G_{ij} \frac{k_0}{T_x b_0}. \tag{9.12}$$

After the setting-up, we have to rearrange both matrices according to the boundary conditions and afterwards we can set up the final matrix of the system of linear equations (see Table 9.13). Thereby we gained the left side of the final system in the matrix **G** and the right side of the vector **DRS**.

The right side is further derived to include the effects of point sources or sinks with the help of formulas from Chapter 5.1.4.8 (see Tab 9.14).

Here the setting-up of the system ends. The solution is carried out by a subprogram SolveEqu which uses the Gauss elimination method with a partial

Table 9.12. Integration and a setup of the matrices **G** and **H**

```
for (j=0;j<nel;j++)
{
    ij=ind (i,j,nel);h[ij]=g[ij]=0.0;
}
for (i=0;i<nel;i++)
{
    CElement* el=GetElement(i);
    i1=el->nod1;i2=el->nod2;
    CNode* nd1=GetNode(i1);CNode* nd2=GetNode(i2);
    xm=(nd1->xb+nd2->xb)*0.5;ym=(nd1->yb+nd2->yb)*0.5*txy;
    hh=0.0;
    for (j=0;j<nel;j++)
    {
        CElement* elj=GetElement(j);
        i1=elj->nod1;i2=elj->nod2;
        CNode* ndj1=GetNode(i1);CNode* ndj2=GetNode(i2);
        x1=ndj1->xb;y1=ndj1->yb*txy;
        x2=ndj2->xb;y2=ndj2->yb*txy;
        ij=ind (i,j,nel);
        Integ (xm,ym,x1,y1,x2,y2,hp,gp,tt);
        if(i!=j) hh+=hp;
        h[ij]=hp+drs[j]*gp;
        g[ij]=gp;
    }
    h[ind(i,i,nel)]-=hh; //diagonal member H [i,i]
}
```

pivoting. Under partial pivoting we understand that the elimination is done in the column that has the biggest absolute value in the eliminated equation. The order of elimination of equations does not change.

9.1.2.2
Commands of Program

The program BemSolve uses a so-called single document ordering which means the program can access only one document at a time (in contrast to BemInp). The main window is a little simpler. It is again composed of the main menu, the tool bar and the status bar. The main menu commands are clearly displayed in Table 9.15.

9.1.2.3
Commands of File Menu

Every command from this menu is identical with the one from a similar menu in the BemInp program, but in contrast to BemInp, this program has to begin with an existing document and so the command *New* is missing.

Table 9.13. Create a global matrix of the system of linear equations

```
for (j=0;j<nel;j++)
{
    CElement* el=GetElement(j);
    if (el–>type>1)
    {
        for (i=0;i<nel;i++)
        {
            ii=ind (i,j,nel);
            ch=g[ii];g[ii]=–h[ii];h[ii]=–ch;
        }
    }
}
for (i=0;i<nel;i++)
{
    drs[i]=0.0;
    for (j=0;j<nel;j++)
    {
        ij=ind(i,j,nel);
        drs[i]+=h[ij]*dl[j];
    }
}
```

Table 9.14. Influence of point sources

```
if (nwl>0)
{
    for (i=0;i<nel;i++) // loop of elements
    {
        CElement* el=GetElement(i);
        i1=el–>nod1;i2=el–>nod2;
        CNode* nd1=GetNode(i1);CNode* nd2=GetNode(i2);
        xm=(nd1–>xb+nd2–>xb)*0.5; ym=(nd1–>yb+nd2–>yb)*0.5*txy;
        for (j=0;j<nwl;j++) // loop of wells
        {
            CWell* wl=GetWell(j);
            dx=xm-wl–>xz;dy=ym-wl–>yz*txy;
            qq=wl–>yield/tx;
            ra=1.0/sqrt (dx*dx+dy*dy);
            drs[i]+=qq*log(ra);
        }
    }
}
```

Table 9.15. Commands of the main menu of program BemSolve

Command	Function
File	Opens a pulldown menu that allows you to work with disk files that contain input data. It also contains commands that print the results of solution
Solve	Commands that direct the solution
View	Commands that change the view of data displayed on the screen
Help	Provides help for the user

Table 9.16. Commands of *Solve* menu (program BemSolve)

Command	Function
Equation	Starts setting-up the matrix of the system of linear equations and its solution. After the solution the results are stored in boundary elements (class CElement)
Int.points	Determines the potential and the components of velocity in all points inside the domain
Wells	Determines the potential in all point sources

Table 9.17. Commands of *View* menu (program BemSolve)

Command		Function
Toolbar		Toggles the toolbar on or off
Status bar		Toggles the status bar on or off
Input data	Elements	Displays input data in boundary elements
	Wells	Displays input data in wells
Results	Elements	Displays results in boundary elements
	Internal points	Displays results in internal points
	Wells	Displays results in wells

9.1.2.4
Commands of Solve Menu

The *Solve* command results in a pulldown menu with functions that run the solution process. All of them are in Table 9.16. First, we have to solve the basic equations (function *Equation*). After determining the potential and its derivative on the boundary, we can determine them in inner points (function *Inner points*) and in point sources (function *Wells*).

9.1.2.5
Commands of View Menu

The menu contains commands that change the way objects are displayed on the screen (see Table 9.17). The first two commands are the same as in the BemInp (Section 9.1.1.6).

The next group of commands serves to display the input data of the boundary elements or the point sources. The data displayed for an element are coordinates of the beginning and the end of the element and the kind of boundary condition. The data of a source that are displayed are its name, coordinates, yield and the diameter of the source.

The last group of commands displays the results of the solution. One can view the results in elements, inner points and wells. The results in an element are the potential, its derivative and a flux through this element. In inner points the results are only the potential and the velocity components and in sources the result is the potential. These results can be printed on a printer with the help of the *Print* command in the *File* menu.

9.1.3
Program BemIsol

This program plots contours of the potential after input of data by the BemInp and after the solution by BemSolve. The window of the program consists of standard components such as the main menu, tool bar and the status bar. Here one can open more than one document. The program is guided through by commands from the main menu and other menus.

9.1.3.1
Basic Relations

The interpolation of contours is done in a rectangular net. This net is placed on the domain and the potential is determined in mesh points inside the domain.

$$u_k = \sum_{j=1}^{N} \int_{\Gamma_j} q w_{kj} d\Gamma_j - \sum_{j=1}^{N} \int_{\Gamma_j} u \frac{\partial w_{kj}}{\partial v} d\Gamma_j + \sum_{l=1}^{N_Q} \frac{Q_l}{T_x} w_{kl}, \tag{9.13}$$

where w_{kj} are values of weight functions in the mesh point k and the element Γ_j. The last term on the right side of the equation stands for the influence of point sources (see Chap. 5.1.4). As the BEFLOW system uses elements with constant base functions, Eq. (9.13) can be derived to

$$u_k = \sum_{j=1}^{N} q_j . G_{kj} - \sum_{j=1}^{N} u_j . \tilde{H}_{kj} + \sum_{l=1}^{N_Q} \frac{Q_l}{T_x} . w_{kl}, \tag{9.14}$$

where G and $\tilde{H}$ are the matrices derived in Chapter 5.1.4. and determined by formulas (9.6) to (9.9). To calculate their terms the BemIsol uses the same subprograms as the BemSolve (Sect. 9.1.2). The mesh points outside the domain are not used because there is no potential defined there.

Thereafter, we test every cell of the net if the potential is defined in all four vertices. If this conditions is fulfilled, the potential in the rectangle can be interpolated by means of a bilinear polynomial

$$u(x,y) = a_0 + a_1 x + a_2 y + a_3 xy. \tag{9.15}$$

This polynomial has four unknown coefficients a_i that are set with the help of known values of the polynomial in the vertices (potential values). The formula of a contour of the potential's value I_u has this form

$$a_0 + a_1 x + a_2 y + a_3 xy = I_u. \tag{9.16}$$

This formula is valid only in the rectangle, and the polynomial we use causes the continuity of contours between the rectangles. The derivatives are not continuous and the contours are not regulated curves. The derivatives would be continuous if we used a bicubic polynomial but it would complicate the solution.

9.1.3.2
Commands of Program

The BemIsol program is designed similarly to the BemInp as a multidocument program. This means one can open several documents at a time. The program is controlled from the main menu with its commands in Table 9.18. Every command in the main menu results in a pulldown menu with further functions.

9.1.3.3
Commands of File Menu

The menu has the same group of functions as the program BemSolve. There is no *New* command since the program's purpose is to work with existing documents. Apart from those commands there are *Open Contours* and *Save Contours* that read data about contours from a disk file or store this data in a disk file respectively.

9.1.3.4
Commands of Interpolation Menu

All these functions (Table 9.20) govern the interpolation of contours. The pulldown menu contains only two functions that serve to:

– set up a rectangular grid needed for the interpolation. There are three kinds of a grid; a thin one (30 × 30 pixels), a normal one (40 × 40 pixels) and a dense one (50 × 50 pixels).

Table 9.18. Commands of the main menu of program BemIsol

Command	Function
File	Opens a pulldown menu that allows you to work with disk files that contain input data. It also contains commands that print the contours on the printer
Interpolation	Does the interpolation of contours
Format	Commands for formatting the output (labels, fonts)
View	Commands that change the view of data displayed on the screen
Window	These commands enable you to arrange a multiple-view of multiple documents in the BemIsol window
Help	Provides help for the user

Table 9.19. Commands of *File* menu (program BemIsol)

Command	Function
Open	Opens an existing document that is stored in the disk file
Save	Saves data about a created net of elements to the same disk file
Save as	Saves data about a created net of elements to a specified disk file
Open contours	Opens an ISO-file with data about contours. This file has to be created by Save contours functions
Save contours	Saves current contours to an ISO-file. This file is binary and can be opened only by the command Open contours
Print	Prints the net of contours on the printer
Print preview	Displays the net of contours on the screen as it would appear printed
Print setup	Selects a printer and a printer connection
Exit	Exits BEMISOL

Table 9.20. Commands of *Interpolation* menu

Command		Function
Net	Thin	A thin net (30 × 30 points), determines potential values in mesh points
	Normal	A normal net (40 × 40 points), determines potential values in mesh points
	Dense	A dense net (50 × 50 points), determines potential values in mesh points
Interval	Regular	A regular interval with a given step
	Irregular	An irregular interval of contours

– set an interval of the contours. It can be either regular or irregular. If you want a regular interval, a dialog box opens where you give a maximal and a minimal value of the potential of a contour and its step. The program then creates a regular net of equipotentials of your choice. If you want an irregular interval, the program opens a dialog box (see Fig. 9.5) where there are all the values of contours and you can add new values by pushing the button *Add* or remove them by pushing *Delete*. The irregular interval can be used in addition to a regular net to add contours to areas of interest.

9.1.3.5
Commands of Format Menu

This menu has only two secondary functions. The first one labels the contours and the second one changes the font of the labelling. It is done using a mouse. The command *Label* only turns on or off the labelling. If the function is on, there is a tick by the *Label* command and after a click of the mouse on a point of a contour, the contour is labelled precisely at that point. If there already is a label, it is deleted by this click.

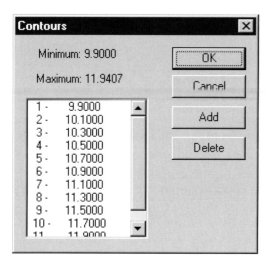

Fig. 9.5. Dialog box for irregular intervals

The function *Font* results in a standard dialog box that allows you to change the font.

9.1.3.6
Commands of Menu View and Window

Both these menus contain commands identical with those in the BemInp program (see Sect. 9.1.1.6).

9.1.4
Program BemStream

The BemStream program calculates and plots the streamlines, but first one has to prepare the input data by the BemInp and run the BemSolve program to solve the system of equations. Afterwards, you can use this program to plot the streamlines. A streamlines is an enveloping curve of the velocity vector and because the program works with a steady flow, the streamlines are identical with trajectories. It is again a multidocument program allowing more than one documents to be opened at the same time. The window of the program works in the same way as the window of the BemInp or the BemSolve and consists of the same components (tool bar with icons, status bar and main menu).

9.1.4.1
Basic Relations

The BEM enables a calculation of the velocity of the flow in an arbitrary point inside the domain using this formula

$$V_{pxk} = \frac{1}{m_e}\left[\sum_{j=1}^{N}\int_{\Gamma_j} q\frac{\partial w_{kj}}{\partial x}d\Gamma_j - \sum_{j=1}^{N}\int_{\Gamma_j} u\frac{\partial}{\partial x}\left(\frac{\partial w_{kj}}{\partial v}\right)d\Gamma_j + \sum_{l=1}^{N_Q}\frac{Q_l}{T_x}\frac{\partial w_{kl}}{\partial x}\right]$$

$$V_{pxy} = \frac{1}{m_e}\left[\sum_{j=1}^{N}\int_{\Gamma_j} q\frac{\partial w_{kj}}{\partial y}d\Gamma_j - \sum_{j=1}^{N}\int_{\Gamma_j} u\frac{\partial}{\partial y}\left(\frac{\partial w_{kj}}{\partial v}\right)d\Gamma_j + \sum_{l=1}^{N_Q}\frac{Q_l}{T_x}\frac{\partial w_{kl}}{\partial y}\right].$$

(9.17)

Equation (9.17) serves to determine the porous velocity at a point k inside the domain. The derivatives of weight functions in a planar case have this form

$$\frac{\partial w_{kj}}{\partial x} = -\frac{x_k - x_j}{r_{kj}^2} \quad \frac{\partial}{\partial x}\left(\frac{\partial w_{kj}}{\partial v}\right) = -2D_{kj}\frac{x_k x_j}{r_{kj}^4},$$

(9.18)

where D_{kj} is an orthogonal distance between the point k and the element j and r_{kj} is a distance between the point k and points of the element j.

The fact that the BEFLOW system uses constant base functions, allows a simple analytic solution of the integrals in Eq. (9.17) analogous to the solution of the integrals in Eq. (9.2). The integration over every element is done by a subprogram InVel in a class CBemStreamDoc (see Table 9.21). One has to run this subprogram for every element. The velocity calculation in a point with coordinates x_b, y_b is done in the class CBemStreamDoc by a Veloc subprogram (see Table 9.22).

So as to construct a streamline, we need to know the beginning point with its coordinates x_0, y_0. Then we will determine the velocity in this point and construct a new point with coordinates x_1, y_1 using the following equations

$$x_1 = x_0 + \Delta\frac{V_{px}}{V} \quad y_1 = y_0 + \Delta\frac{V_{py}}{V},$$

(9.19)

where Δ is a chosen small part of a streamline. It is set as two thousandth of a window's diagonal. The process is repeated again and again and we obtain other points of the streamline until the streamline does not leave the domain.

Moreover, the program allows a bunch of streamlines to be created that begin in a point source. In this case, the starting point of a streamline is the point which lies on a perimeter of the source. The beginning points are placed regularly on the perimeter of the source and their number can be given. The program temporarily reverses the orientation of the velocity vector because the streamlines should end and not begin in the source. The program even enables putting time marks on a streamline that show how far a particle has traveled after a given time.

9.1.4.2
Commands of Program

The BemStream program works the in same way as the BemIsol. The main menu has six functions that are in Table 9.23. Each of these functions result in

Table 9.21. Subprogram InVel for an analytical integration

```
void CBemStreamDoc::InVel(double xp,double yp,double x1,double y1,double x2,double
y2,double& hx,double& hy,double& gx,double& gy,double tt)
{
    double r1,r2,dl,lg1,lg2,nx,ny,d,xob,xop,yop,a2,b2,c2,fa,fb,
        yo1,yo2,dx1,dx2,dy2,dy1,s2,pi=3.14159265,pi2;
    pi2=pi*0.5;dx1=x1-x2;dy1=y1-y2;
    c2=dx1*dx1+dy1*dy1;
    dx1=x1-xp;dx2=x2-xp;dy1=y1-yp;dy2=y2-yp;
    a2=dx1*dx1+dy1*dy1;b2=dx2*dx2+dy2*dy2;
    r1=sqrt(a2);r2=sqrt(b2);
    dl=sqrt(c2);                            // length of the element
    lg1=log(r1);lg2=log(r2);
    ny=(x1-x2)/dl;nx=(y2-y1)/dl;            // cosines of outward normal
    xo2=x2*nx+y2*ny;xop=xp*nx+yp*ny;
    d=xo2-xop;                              // distance
    yo1=-ny*x1+nx*y1;
    yo2=-ny*x2+nx*y2;yop=-ny*xp+nx*yp;
    if (fabs(d)<1e-8)                       // point lies on the line
    {
        gx=(lg2-lg1)*ny;
        gy=-nx*(lg2-lg1);
        hx=hy=0.0;
        gx*=tt;gy*=tt;
    }
    else
    {
        fb=atan2 (yo2-yop,d);fa=atan2 (yo1-yop,d);
        if (fabs(fa)==pi2||fabs(fb)==pi2)
        {
            gx=(lg2-lg1)*ny;
            gy=-nx*(lg2-lg1);
            hx=hy=0.0;
            gx*=tt;gy*=tt;
        }
        else
        {
            s2=fa-fb;
            gx=nx*s2+ny*(lg2-lg1);
            gy=ny*s2-nx*(lg2-lg1);
            hx=dy2/b2-dy1/a2;
            hy=-dx2/b2+dx1/a2;
            hx*=tt;hy*=tt;
            gx*=tt;gy*=tt;
        }
    }
}
```

Table 9.22. Subprogram Veloc for computation of velocity components

```
void CBemStreamDoc::Veloc (double& vx,double& vy,float xb,float yb)
{
   int j,ip,i1,i2;
   float x1,y1,x2,y2,qq,ddf,sx,sy,dx,dy,rb,ra,moc1,ybb;
   double ax,ay,bx,by;
   tx=kx*thick;ty=ky*thick;
   tt=sqrt(ty/tx);              // Jacobian of transformation
   tp=1.0/(6.2831852*tt);
   txy=sqrt(tx/ty);             // transformation of y-coordinates
   ybb=yb*txy;
   vx=vy=0.0;
   for (j=0;j<nel;j++)          // integration loop
   {
      CElement* el=GetElement(j);
      i1=el->nod1;i2=el->nod2;
      CNode* nd1=GetNode(i1);CNode* nd2=GetNode(i2);
      x1=nd1->xb;y1=nd1->yb*txy;
      x2=nd2->xb;y2=nd2->yb*txy;
      InVel (xb,ybb,x1,y1,x2,y2,ax,ay,bx,by,tt);
      vx+=el->der*bx-el->pot*ax;
      vy+=el->der*by-el->pot*ay;
   }
   if (nwl>0)                   // influence of wells
   {
      sx=sy=0.0;
      for (j=0;j<nwl;j++)
      {
         CWell* wl=GetWell(j);
   dx=xb-wl->xz;dy=ybb-wl->yz*txy;
   rb=dx*dx+dy*dy;
         qq=-wl->yield/tx;
   ra=1./sqrt(rb);
   sx+=qq*dx/rb;sy+=qq*dy/rb;
      }
      vx+=sx;vy+=sy;
   }
   vx*=tp*kx;vy*=tp*ky*txy;  // transformations
}
```

Table 9.23. Commands of the main menu of BemStream

Command	Function
File	Opens a pulldown menu that allows you to work with disk files that contain input data. It also contains commands that print the streamlines on the printer
Streamlines	Sets up the net of streamlines
Format	Edits the final net of streamlines, sets up the time marks
View	Commands that change the view of data displayed on the screen
Window	These commands enable you to arrange a multiple-view of multiple documents in the BemStream window
Help	Provides help for the user

a pulldown menu. The functions *Windows* and *Help* are identical with those in the BemIsol (see Sect. 9.1.1.7).

9.1.4.3
Commands of File Menu

This menu is the same as in the BemIsol (Sect. 9.1.3.3). Instead of commands *Open* and *Save contours* there are *Open* and *Save streamlines* that read and store streamlines together with the time marks.

9.1.4.4
Commands of Streamlines Menu

The *Streamlines* option has only a menu with two functions that serve to set the beginning of a streamline. Using the *Point* command, the starting point of a streamline is chosen by clicking the mouse's left button in the domain. The second command *From well* opens a dialog box where you have to give the number of streamlines that end in the source. After that you have to select the source using a mouse.

9.1.4.5
Commands of Format Menu

Here you can define the time marks and change their font. Before defining the time marks, it is recommended to activate the *Time* function from the *View* menu (Sect. 9.1.4.6) to know a maximal time of each streamline. It will help you to choose appropriate time intervals of the marks.

After using the function *Time marks*, a dialog box appears (Fig. 9.6) where you can set and edit the marks. Every mark is defined by its time and colour. When you finish and press *OK*, the time marks are drawn in the net.

Table 9.24. Commands of *File* menu (program BemStream)

Command	Function
Open	Opens an existing document that is stored in the disk file
Save	Saves data about a created net of elements to the same disk file
Save as	Saves data about a created net of elements to a specified disk file
Open streamlines	Opens an STR-file with data about streamlines. This file has to be created by Save streamlines functions
Save streamlines	Saves current streamlines to an STR-file. This file is binary and can be opened only by the command Open streamlines
Print	Prints the net of streamlines on the printer
Print preview	Displays the net of streamlines on the screen as it would appear printed
Print setup	Selects a printer and a printer connection
Exit	Exits BEMISOL

Fig. 9.6. Dialog box for input of time marks

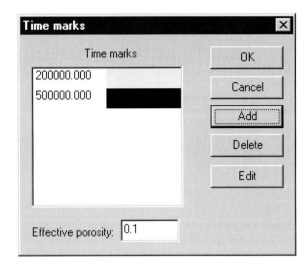

9.1.4.6
Commands of View Menu

Apart from the standard commands from the BemIsol or the BemInp (Sect. 9.1.1.6), there are two more commands here. The first new function *Time* shows or hides the total time which it takes a particle of fluid to go along the whole streamline. This time is written at the end of each streamline and its unit depends on the unit of the coefficient of hydraulic conductivity [coefficient's unit (m s^{-1}) implies the time's unit (s); the coefficient's unit (m day^{-1}) implies the time's unit (day)]. The second function *Time marks* shows or hides the time marks. If we have defined the time marks, the function temporarily hides them.

9.1.5
Program BemRaw

This program combines two methods; the boundary element method to determine the velocity of a flow and the random-walk method to simulate the pollution in groundwater (see Chap. 6.1.5). It is a part of the BEFLOW system because it uses input data of a compatible format. Before running it, one has to run the BemInp to create a net of elements and the BemSolve to solve the system of equations. In contrast to a real program, the result of this program is a cloud of particles. The transformation of particles to a concentration of the pollutant is missing and to do that one has to generate a net to add together the particles in every cell of the net. The program can solve a case with a linear equilibrium sorption including the pollutant's decay.

Table 9.25. Commands of *View* menu (program BemStream)

Command	Function
Toolbar	Shows or hides the toolbar
Status Bar	Shows or hides the status bar
Zoom in	Zooms in the picture on the screen
Zoom out	Zooms out the picture on the screen
Zoom full	Plots the picture in the initial size
Time	Shows or hides the total time along a streamlin
Time marks	Shows or hides time markers

9.1.5.1
Basic Relations

As we have already mentioned, the program uses the boundary element method to determine the velocity of a groundwater flow. So it is obvious that the program has the same subprograms Invel and Veloc as the BemStream (Sect. 9.1.4). There are new classes CPartic to work with particles and CTimeInt for time intervals (the classes' declarations are in Table 9.26). Every time interval contains a field of particles. A particle is defined by its position in the domain (by coordinates x_b, y_b). Every particle has also another set of coordinates x_i, y_i which is its coordinates in the program's window. The initial position of a particle is generated by the program as described below. First, we have to set the polluted area of a polygonal shape by giving the coordinates of its vertices. Then we have to select the density of particles and set all values needed for the transport. The most important values are the longitudinal and the transversal dispersivities a_L, a_T and the coefficient of effective porosity. These values have to be set always. If we consider a linear sorption, we have to set the coefficient of distribution K_d and the soil's bulk density ρ_s. Finally, we have to set the time intervals and thereby we trigger the calculation. The new position of particles is plotted on the screen after every time interval. The core of the calculation is done by the PositionParts subprogram whose most important part can be seen in Table 9.27. The velocity is calculated twice in every time interval, once in the starting position of the particle and once in the final position of the particle. These two velocities are averaged for further use and also allow to estimate a needed transformation (see Chap. 6.1.5) which has this form

$$v'_x = v_x + \frac{\partial D_{xx}}{\partial x} + \frac{\partial D_{xy}}{\partial x} = v_x + \frac{\partial}{\partial x}\left[a_L \frac{v_x^2}{v} + a_T \frac{v_y^2}{v} + (a_L - a_T)\frac{v_x v_y}{v}\right]$$

$$v'_y = v_y + \frac{\partial D_{yy}}{\partial y} + \frac{\partial D_{yx}}{\partial y} = v_y + \frac{\partial}{\partial y}\left[a_L \frac{v_y^2}{v} + a_T \frac{v_x^2}{v} + (a_L - a_T)\frac{v_x v_y}{v}\right],$$

(9.20)

where v_x, v_y are velocity components in the x,y axis direction and v is the length of the vector of velocity.

Table 9.26. Declaration of classes of program BemRaw

```
class CTimeInt:public CObject
{
public:
    CTimeInt (float tim);
    ~CTimeInt();
    CPartic* AddParts(float x,float y);
    CPartic* GetPartic(int i);
    virtual void Serialize (CArchive& ar);
protected:
    CTimeInt();
    DECLARE_SERIAL (CTimeInt);
public:
    float tim;
    CObArray Parts;
};
class CPartic:public CObject
{
public:
    CPartic (float x,float y);
    CPartic (CPoint p,float *gc);
protected:
    CPartic();
    DECLARE_SERIAL (CPartic);
public:
    BOOL operator ==(const CPartic& other) const;
    void TranCoord (float *gcb);
    void RevTran (float *gcb);
    void Coord (float& x,float& y);
    CPoint Coord(void);
public:
    float xb,yb;
    int xi,yi;
public:
    void Plot (CDC& dc, float *gc, COLORREF clr);
    virtual void Serialize (CArchive& ar);
};
```

9.1.5.2
Commands of the Program

BemRaw has the main menu with six basic functions (see Table 9.28). Most of them are similar to the functions of BemInp. The program also uses a multiple document template and can work with more documents at the same time. The *File* command works with files and printers as always. It allows you to write or read data about time intervals and particles which are stored in files with an extension PAR. These are binary files that are created with the help of the Serialize function. By the *Geometry* command you can set a geometric shape

Table 9.27. Computer code used to compute the position of particles

```
    retd=efpo+rho*kd;                            // retardation factor
    for (it=0;it<ntim;it++)       //--- time loop ---------
    {
        CTimeInt* tm1=GetTime(it);
        tim+=delt;
        CTimeInt* tm2=AddTime(tim);
        npar=tm1->Parts.GetSize();
        for (ipa=0;ipa<npar;ipa++)    // particles loop
        {
            CPartic* par=tm1->GetPartic(ipa);
            xb=par->xb;yb=par->yb;
            Veloc (vx,vy,xb,yb);       // computation of velocity
            vx/=retd;vy/=retd;
            dx=vx*delt;dy=vy*delt;
            xb+=dx;yb+=dy;
            pt=TranCoord(xb,yb);
            if (rg.PtInRegion(pt))                     //test, if a particle is into the area
            {
                Veloc (vxx,vyy,xb,yb);     //velocity in a new position
                vxx/=retd;vyy/=retd;
                v=sqrt(vx*vx+vy*vy);   // transformation of velocity
                dvx=vxx-vx;dvy=vyy-vy;dv=sqrt(vxx*vxx+vyy*vyy)-v;
                dxx=(al*(2.0*vx*dvx*v-vx*vx*dv)+
                    at*(2.0*vy*dvy*v-vy*vy*dv))/(v*v*dx);
                dyy=(al*(2.0*vy*dvy*v-vy*vy*dv)+
                    at*(2.0*vx*dvx*v-vx*vx*dv))/(v*v*dy);
                dxy=(al-at)*((dvx*vy+dvy*vx)*v-vx*vy*dv)/(v*v*dx);
                dyx=(al-at)*((dvx*vy+dvy*vx)*v-vx*vy*dv)/(v*v*dy);
                vx=(vxx+vx)*0.5+dxx+dxy;vy=(vyy+vy)*0.5+dyy+dyx;
                v=sqrt(vx*vx+vy*vy);r1=r2=0.0F;
                if (al>0.0F)                // longitudinal disperzivity
                {
                    sigl=sqrt(2*al*v*delt)/v;r1=gran();
                }
                if (at>0.0F)                // transversal disperzivity
                {
                    sigt=sqrt(2*at*v*delt)/v;r2=gran();
                }
                dx=vx*(delt+r1*sigl)+vy*r2*sigt;
                dy=vy*(delt+r1*sigl)+vx*r2*sigt;
                xx=par->xb+dx;yy=par->yb+dy;     //--- new position
                CPartic* prn=tm2->AddParts(xx,yy);
                prn->Plot(td,gcb,RGB(ired,igren,ibl));
            }
        }                         //------ end of particles loop
    }                             //------ end of time loop
```

9.1 BEFLOW System

Table 9.28. Commands of the main menu of program BemRaw

Statement	Function
File	Opens a pulldown menu that allows you to work with disk files that contain input data. It also contains commands that print the cloud of particles on the printer
Geometry	Defines the polluted area and generates particles
Time intervals	Sets time intervals and starts a simulation
View	Commands that change the view of data displayed on the screen
Window	These commands enable you to arrange a multiple-view of multiple documents in the BemRaw window
Help	Provides help for the user

Table 9.29. Commands of *File* menu (program BemRaw)

Command	Function
Open	Opens an existing document that is stored in the disk file
Save	Saves data about a created net of elements to the same disk file
Save as	Saves data about a created net of elements to a specified disk file
Open particles	Opens a PAR-file with data about particles. This file has to be created by Save particles functions
Save particles	Saves current set of particles to a PAR-file. This file is binary and can be opened only by the command Open particles
Print	Prints the cloud of particles on the printer
Print preview	Displays the cloud of particles on the screen as it would appear printed
Print setup	Selects a printer and a printer connection
Exit	Exits BEMRAW

of the initial polluted area along with data required to generate particles. Time intervals can be defined by the command of the same name. After having finished, the program starts the calculation of the pollution's transport. The remaining functions have already been discussed.

9.1.5.3
Commands of File Menu

As all the commands of the *File* menu are similar to those used in other programs, they are only reviewed in Table 9.29.

9.1.5.4
Commands of Geometry Menu

The *Geometry* menu has two functions:

Area – defines the polluted area. It opens a cascaded menu where you can choose whether you want to give the vertices of the polygon (polluted area) or you want to load the coordinates from a TXT file or save an already given set of coordinates into a TXT file. Commands that set the vertices are analogous to the commands that set nodes of a domain in the BemInp program (Sect. 9.1.1.2).

Particles – here a dialog box is opened where the number of particles in the x and y axis direction is given. The program generates a rectangular net that covers the polluted area. The calculation considers only particles that are inside this area. This implies that one has to define the polluted area first and then one can generate particles. Beside the density of particles, one has to set the coefficients of longitudinal and transversal dispersivity, the coefficient of effective porosity and (optional) the coefficient of distribution and the bulk density of the soil's skeleton in the same dialog box.

9.1.5.5
Command Time Intervals

In contrast to other commands this one does not open any pulldown menu but a dialog box where the number of time steps and the interval of one step should be set. The BemRaw uses time steps of the same length. After pressing the *OK* button, the program starts the simulation. It is obvious that the time intervals have to be set after a definition of the polluted area and after generating the particles in the *Geometry* menu. New particles are created in every time step and this causes high memory requirements in tasks with a large number of particles or time steps.

9.1.5.6
Commands of View Menu

Here, only one function is different to functions in the BemInp's *View* menu (Sect. 9.1.1.6). The new function shows particles only in selected time intervals by means of a dialog box which serves to give an initial and a final time interval that should be showed. A default settings show all time intervals. Other functions are identical with the aforementioned programs (see Table 9.30).

Table 9.30. Commands of *View* menu (program Bemraw)

Command	Function
Toolbar	Shows or hides the toolbar
Status Bar	Shows or hides the status bar
Zoom in	Zooms in the picture on the screen
Zoom out	Zooms out the picture on the screen
Zoom full	Plots the picture in the initial size
Time	Shows particles from selected time interval

9.2
System UNSDIS

The system UNSDIS uses the finite element method for one-dimensional models of a groundwater flow in an unsaturated zone together with the pollution's transport (see Chaps. 5.2 and 6.2). The transport's solution includes a case of an equilibrium sorption with the Freundlich isotherm (Chap. 3.1.4). The groundwater flow simulation contains parts of a code (Khaleel and Yeh 1985) used with a kind permission of the National GroundWater Association, Copyright 1985.

9.2.1
Basic Relations

A pressure potential is used in models of an unsaturated zone (in contrast to the saturated zone where a sum of pressure and gravity potential is used, see Chap. 2.2). The pressure potential is represented by a pressure head and it equals zero at the groundwater level. It is negative in the unsaturated zone and positive in the saturated zone. A relationship between the coefficient of unsaturated conductivity and the pressure potential is (see van Genuchten 1980 or Chap. 2.3)

$$K(H) = k_s \frac{\left\{1 - (\alpha|H|)^{n-1}\left[1 + (\alpha|H|)^n\right]^{-m}\right\}^2}{\left[1 + (\alpha|H|)^n\right]^{\frac{m}{2}}}. \tag{9.21}$$

The relationship between the moisture and the potential has this form

$$\Theta(H) = \Theta_r + (\Theta_s - \Theta_r)\left[1 + (\alpha|H|)^n\right]^{-m} \tag{9.22}$$

Here we used

Θ_s – soil's moisture by a total saturation,
Θ_r – residual soil's moisture,
k_s – coefficient of hydraulic conductivity for a fully saturated medium,
H – pressure potential of the flow (pressure head),
α, n – coefficients, $m = 1 - \dfrac{1}{n}$

This program is written similarly to the BEFLOW system in the Visual C++ programming language, version 6.0 and it uses the simplest linear element (see Chap. 5.2.3) in a one-dimensional model of a vertical flow. Main classes are (see also Table 9.31):

- *CLayer* – a class describing properties of soil layers. Each layer is given by an initial and a final depth under the soil's surface. All coefficients are

Table 9.31. Declaration of classes of program UNSDIS

```
class CLayer:public CObject
{
protected:
    CLayer();
    DECLARE_SERIAL (CLayer);
public:
    CLayer (float z1,float z2,int nel,float ths,float thr,float n,
            float alf, float ks);
    float Theta (float psi);
    float Kuns (float psi);
    float Capac (float psi);
    float Diffus (float psi);
    float Se (float psi);
    virtual void Serialize (CArchive& ar);
    float alpha,wn,ths,thr,ksat;
    float disp,ro,stn,kd,efpo,potin,conin;
    float zbeg,zend;
    int nel;
};
class CElement:public CObject
{
protected:
    CElement();
    DECLARE_SERIAL (CElement);
public:
    CElement (float z1,float z2);
    float leng();
    virtual void Serialize (CArchive& ar);
    float z1,z2;
    int nlay;
};
class CResult:public CObject
{
protected:
    CResult();
    DECLARE_SERIAL (CResult);
public:
    CResult (float ps);
    virtual void Serialize (CArchive& ar);
    float psi, theta, conc, cons;
};
class CTimeStep:public CObject
{
protected:
    CTimeStep();
    DECLARE_SERIAL (CTimeStep);
public:
    CTimeStep (float tim);
    ~CTimeStep();
    CResult* GetResult (int i);
    virtual void Serialize (CArchive& ar);
    CObArray Results;
    float time,pmax,cmax,pmin,cmin;
};
```

constant in one layer and this class contains all the coefficients that characterise the flow in the unsaturated zone Θ_s, Θ_r, k_s, α and n [see Eqs. (9.21) and (9.22)]. Moreover, there are characteristic coefficients of the transport of pollution a_L, a_T, K_d, ρ and the Freundlich coefficient n_F. Apart from this, the class contains information about the initial values of the pressure head and the concentration of pollution in the solution. The distance between the two depths is divided into a given number of parts (of the same length) that represent finite elements. The class also contains functions to determine the moisture (Theta), the coefficient of unsaturated conductivity (Kuns), the coefficient of moisture capacity (Capac) and the coefficient of molecular diffusion (Diffus) and their relationship to the pressure potential.
- *CElement* – a class representing a finite element. It is a one-dimensional element and the class stores the coordinates of an initial and a final node and the number of the layer.
- *CResult* – a class containing all kinds of results. These results include the pressure potential (psi), the moisture (theta), the concentration of the pollutant in the solution (Conc) and the concentration of the pollutant sorbed on the soil's skeleton (Cons).
- *CTimeStep* – a class of time steps. This class stores results in every time step. It contains the time and an array of results at this time. To plot these results in a graph, the class stores a minimal and a maximal value of the potential and the concentration in this time step.

Beside these classes, the program uses standard classes generated by the Visual C++ with the most important classes being derived from the CDocument and CView base classes from the MFC library (see Kruglinski 1998, Prosise 1996). To display the results we again use the CZoomView (Pirtle 1995).

9.2.2
Commands of Program

The window of the program looks just like the windows of the BEFLOW system with a main menu, a toolbar and a status bar. Here one can open more documents at the same time as the program is based on the multidocument template. Commands of the main menu are in Table 9.32 and most of them result in a pulldown menu.

9.2.2.1
Commands of the File Menu

This is a standard menu as in the BemInp program (Sect. 9.1.1.2). It works with disk files and prints documents on the printer. The disk files used here have a UNS extensions and they can be accessed only by this system. The last command *Exit* finishes the work with this system.

The only new command comparing to BemInp is the function *Save Results* that saves results in every time step into a text file and they can be imported

Table 9.32. Commands of the main menu of program UNSDIS

Command	Function
File	Opens a pulldown menu that allows you to work with disk files that contain input data. It also contains commands that print the results on the printer
Input data	Creates a new set of input data or edits an existing set
Solve	Starts a simulation
Format	Edits the output
View	Commands that change the view of data displayed on the screen
Window	These commands enable you to arrange a multiple-view of multiple documents in the UNSDIS window
Help	Provides help for the user

into different text processing programs. The data that are saved in every time step are the node's number, its depth under the soil's surface, the pressure head, the moisture and the pollutant's concentrations in the solution and sorbed on the soil's skeleton.

9.2.2.2
Commands of Input Data

This group of commands prepares a set of input data or edits an already existing set. It is comprised of these three functions:

- *Layers* This option allows to create a new or edit or delete an existing layer. The statement *Add* opens a dialog box to define a layer where one has to give the depth of the shallowest part and of the deepest part of the layer in centimetres and the number of elements which will be in the layer. It is further important to set all coefficients needed for a solution (see Sect. 9.2.1) together with values of an initial pressure head (cm) and the concentration. The coefficient of hydraulic conductivity k_s is given in (cm day^{-1}) and the coefficient of distribution and the bulk density must have corresponding dimensions; for example K_d(m^{-3} kg^{-1}) and ρ(kg^{-1} m^3). Other coefficients are non-dimensional.
- *Boundary* opens a dialog box where boundary conditions on the top and on the bottom boundary of the domain are set.
- *Time* results in a dialog box that sets the time intervals. The unit of intervals is one day. The program uses two kinds of time intervals. One must give only the intervals where one wants to know the results so at least one such interval must be given. The program uses secondary intervals which it generates according to the data you put in and according to the result in the previous interval (see Khaleel and Yeh 1985). The needed data are also put in here:
- *Initial time step* is given here and it is used when generating the secondary intervals. It is given in days and should be sufficiently small.

- *Maximal step length.* This restriction from above prevents a big increase in the length of secondary intervals.
- *Tolerance.* This sets a maximal deviation of the pressure head in the entire domain. If this restriction is exceeded, the iterative solution stops and the program continues with the next time interval.
- *Max. number of iterations.* This condition stops a possible slow convergence of the iteration.
- *Time schema.* Here the ξ-parameter from Eq. (5.146) can be set and it determines the type of a schema. A recommended value is 1 (a fully implicit schema).

9.2.2.3
Command Solve

Hereby you start the solution. In the course of the solution a graph of the pressure head and of the concentration is plotted on the screen in selected time intervals.

9.2.2.4
Commands of Format Menu

These commands direct the layout on the screen or on the printer.

- *Axes* opens a dialog box where you can give a range of the axes in graphs of the pressure head and the concentration by giving a minimal and a maximal value. Apart from that, you can give a step of the tics or a step of the labels on the axes.
- *Font.* Here you can choose a font and its size.
- *Measure.* The default measure is 1:100 but here it can be changed.

9.2.2.5
Commands of View Menu

Most of these commands are the same as in the program BemInp (Sect. 9.1.1.6) and in addition to them there is a command that displays the results only in

Table 9.33. Commands of *Input data* menu

Command		Functions
Layers	Add	Creates a new layer
	Delete	Removes a chosen layer
	Edit	Corrects data of a selected layer
	Generate	Generates finite elements after input or correction of dat
Boundary		Sets the boundary conditions
Time		Sets the time intervals

Table 9.34. Commands of *View* menu (program UNSDIS)

Command	Function
Toolbar	Shows or hides the toolbar
Status Bar	Shows or hides the status bar
Zoom in	Zooms in the picture on the screen
Zoom out	Zooms out the picture on the screen
Zoom full	Plots the picture in the initial size
Time ...	Shows only selected time intervals
Graph	Switches into a graphical form of display of results
Tabular	Switches into a numerical form of display of results

selected time intervals. The only thing you have to give is the first and the last time interval that should be displayed. Besides here you can opt for two ways of displaying the results either in the form of a graph or a table. Due to the fact that this is not a full version, the output is limited only to one page. All commands can be seen in Table 9.34.

9.3
Examples

These simple examples are on the CD only to try out the software and all the commands described above.

9.3.1
System BEFLOW

9.3.1.1
Example Square 1

This is a simple square domain with a size of 100×100 m, the coefficient of hydraulic conductivity is $k = 1 \times 10^{-4}$ m s^{-1} and the thickness of an aquifer is 10 m. The boundary conditions on the two horizontal sides of the square are $q = 0$ (it is an impermeable boundary). The remaining boundary conditions set the potential on the left side to $H = 12$ m and on the right to $H = 10$ m. The water flows from left to right. Every side of the domain is divided into 10 elements, thus the total number of elements is 40.

It is obvious that the piezometric level in a steady state is a plane which has a slope of 2%. This allows us to check the values of the derivative that should equal 0.02 on the vertical sides. A comparison of the model's result with an exact solution is in the Table 9.35.

Streamlines are horizontal lines and equipotentials are orthogonal to them and form a set of vertical lines. This example clearly shows an error in the cal-

Table 9.35. Comparison of results of example Square1 with exact solution

Potential				Derivative			
Element	Model	Exact	Error	Element	Model	Exact	Error
1	11.9056	11.9	−0.0056	11	0.02099	0.02	−0.00099
2	11.7033	11.7	−0.0033	12	0.01965	0.02	0.00035
3	11.502	11.5	−0.002	13	0.01991	0.02	9E-05
4	11.3011	11.3	−0.0011	14	0.01994	0.02	6E-05
5	11.1004	11.1	−0.0004	15	0.01996	0.02	4E-05
6	10.8996	10.9	0.0004	16	0.01996	0.02	4E-05
7	10.6989	10.7	0.0011	17	0.01994	0.02	6E-05
8	10.498	10.5	0.002	18	0.01991	0.02	9E-05
9	10.2967	10.3	0.0033	19	0.01965	0.02	0.00035
10	10.0944	10.1	0.0056	20	0.02099	0.02	−0.00099

culation of velocities that occurs near the impervious boundary. The streamlines should be parallel with this boundary though in a close neighbourhood they are sinuous due to the error (see Fig. 9.7). An exact solution of the integrals reduces the error but it does not remove it completely.

9.3.1.2
Example Circle1

It is the simplest example of a task with a point source. The domain is a circle with a diameter of 20 m and with a point source W-1 in the centre of the circle. The coefficient of hydraulic conductivity is $k = 1 \times 10^{-4}\,\mathrm{m\,s^{-1}}$ and the thickness of an aquifer is 10 m. The source has a diameter of 0.2 m and a yield of $1 \times 10^{-3}\,\mathrm{m^3\,s^{-1}}$. There are three inner points where the groundwater level is determined and their distance from the centre is 2, 5 and 10 m.

The results of this model can be well compared with an exact solution that satisfies the Dupuit equation

$$\Phi = \frac{Q(\ln r - \ln R)}{2\pi T}, \qquad (9.23)$$

where r is a distance from the source, R is a distance between the boundary condition $h = 0$ and the source and T is the transmissivity coefficient. The comparison is given in Table 9.36. The potential is negative because it is a relative drawdown.

The net of streamlines is simple and can be seen in Fig. 9.8. The equipotentials are concentric circles and the streamlines are lines that stem from the source.

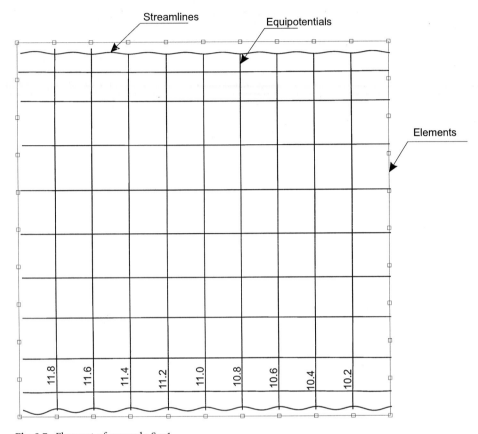

Fig. 9.7. Flow net of example Sqr1

Table 9.36. Results of example Circle1

Point	Distance from well [m]	Model	Exact solution	Differences
1	2	−0.36600	−0.36647	−0.000467
2	5	−0.22010	−0.22064	−0.000535
3	10	−0.10980	−0.11032	−0.000517
Well	0.1	−0.84274	−0.84325	−0.000513

9.3.1.3
Example Square 2

This is similar to the previous example with only one difference of a point source W-1 being in the middle of the domain. The diameter of the source is 0.2 m and its yield is $1 \times 10^{-3}\,\mathrm{m^3\,s^{-1}}$. The net of streamlines is shown in Fig. 9.9.

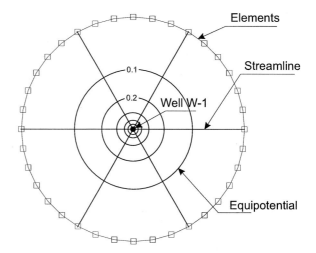

Fig. 9.8. Flow net of example Circ1

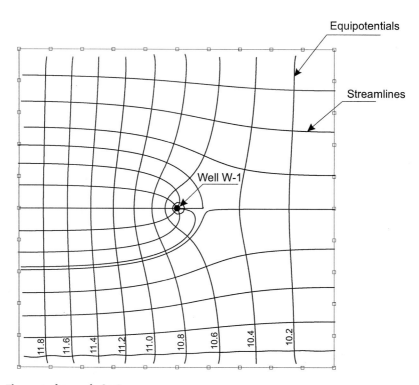

Fig. 9.9. Flow net of example Sqr2

Table 9.37. Input data for examples Uns1 and Uns2

Quantity	Abbrevation	Value	Units
Saturated moisture	Θ_s	0.35	–
Residual moisture	Θ_r	0.02	–
Coefficient of hydraulic conductivity	K_s	0.000722	cm s^{-1}
Coefficient α	α	0.041	–
Coefficient n	n	1.964	–
Initial pressure head	φ_0	−150	cm
Initial concentration	C_0	0	–
Dispersivity	a_L	1	cm

Table 9.38. Results of example Uns1

Time: 1800.000000

No.	Depth	Potential	Moisture	Conc.in fluid	Sorbed
1	0.00	0.750	0.350	100.000	0.000
2	0.25	0.612	0.350	99.839	0.000
3	0.50	0.474	0.350	99.665	0.000
4	0.75	0.336	0.350	99.478	0.000
5	1.00	0.199	0.350	99.277	0.000
6	1.50	−0.081	0.350	98.833	0.000
7	2.00	−0.374	0.350	98.328	0.000
8	3.00	−1.033	0.350	97.102	0.000
9	4.00	−1.807	0.349	95.571	0.000
10	5.00	−2.729	0.348	93.695	0.000
11	6.00	−3.847	0.346	91.432	0.000
12	7.00	−5.237	0.342	88.735	0.000
13	8.00	−7.029	0.337	85.544	0.000
14	9.00	−9.466	0.327	81.752	0.000
15	10.00	−13.081	0.311	77.097	0.000
16	11.00	−19.354	0.279	70.920	0.000
17	12.00	−33.867	0.216	62.042	0.000
18	13.00	−86.548	0.114	44.281	0.000
19	14.00	−147.621	0.077	3.056	0.000
20	15.00	−149.975	0.077	0.089	0.000
21	16.00	−150.000	0.077	0.003	0.000
22	17.00	−150.000	0.077	0.000	0.000
23	18.00	−150.000	0.077	0.000	0.000
24	19.00	−150.000	0.077	0.000	0.000
25	20.00	−150.000	0.077	0.000	0.000

9.3.2
Program UNSDIS

9.3.2.1
Example Uns1

It is not easy to pick an example that can be compared with its solution for the Unsdis system. The one we use is a simplified example from (Vogel, 1987). The report uses a more complicated model of a non-linear relationship between the coefficient of unsaturated conductivity and the pressure potential hence the results cannot be precisely compared. The example is satisfactory for an illustration of the program's work. The data in case of a flow in the unsaturated zone can not be chosen arbitrarily because the solution does not have to converge. The report focuses on the groundwater flow and not on the pollution's transport. The transport will be set without sorption ($K_d = 0$), the boundary condition is set to $C = 100\%$. The boundary condition of the groundwater flow

Table 9.39. Results of example Uns2

Time: 1800.000000

No.	Depth	Potential	Moisture	Conc.in fluid	Sorbed
1	0.00	0.750	0.350	3.500	0.539
2	0.25	0.612	0.350	3.428	0.528
3	0.50	0.474	0.350	3.351	0.516
4	0.75	0.336	0.350	3.269	0.503
5	1.00	0.199	0.350	3.183	0.490
6	1.50	−0.081	0.350	2.995	0.461
7	2.00	−0.374	0.350	2.793	0.430
8	3.00	−1.033	0.350	2.348	0.362
9	4.00	−1.807	0.349	1.889	0.291
10	5.00	−2.729	0.348	1.449	0.223
11	6.00	−3.847	0.346	1.055	0.162
12	7.00	−5.237	0.342	0.727	0.112
13	8.00	−7.029	0.337	0.471	0.072
14	9.00	−9.466	0.327	0.284	0.044
15	10.00	−13.081	0.311	0.156	0.024
16	11.00	−19.354	0.279	0.076	0.012
17	12.00	−33.867	0.216	0.027	0.004
18	13.00	−86.548	0.114	0.003	0.001
19	14.00	−147.621	0.077	0.000	0.000
20	15.00	−149.975	0.077	0.000	0.000
21	16.00	−150.000	0.077	0.000	0.000
22	17.00	−150.000	0.077	0.000	0.000
23	18.00	−150.000	0.077	0.000	0.000
24	19.00	−150.000	0.077	0.000	0.000
25	20.00	−150.000	0.077	0.000	0.000

is a flood above the soil's surface of 0.75 cm and on the bottom it is $q = 0$ (no outflow).

The setting of the time was also adapted from the report where the time intervals are provided in seconds. The calculation was run for 60, 900, 1800, 2700, 3600, and 5400 s. The input data are in Table 9.37 and the results in the selected intervals are in Table 9.38.

9.3.2.2
Example Uns2

This example differs from the Uns1 only in the second part (the one about pollution transport). Now there is a linear sorption and we set $K_d = 0.154\,dm^3\,kg^{-1}$ for every layer. We added a boundary condition of a constant concentration of pollutants in the solution that enters the top of the domain ($C = 3.5\,mg\,l^{-1}$). It is a linear sorption because the exponent is $n_F = 1$. All results can be found in Table 9.39.

References

Khaleel R, Yeh TC (1985) Galerkin-finite element program for simulating unsaturated flow in porous media. Ground Water 23:90–96
Kruglinski DJ (1998) Inside Visual C++. Microsoft Press, Redmont
Pirtle B (1995) Automatic scaling for MFC. Windows Developers J. 6:10
Prosise J (1996) Programming Windows95 with MFC. Microsoft Press, Redmont
van Genuchten MT (1980) A closed-form equation for predicting the hydraulic conductivity of unsaturated soils. Soil Sci. Soc. Am. J. 44:892–898
Vogel T (1987) SWMII – Numerical model of two-dimensional flow in a variably saturated porous medium. Research rep.87, Agric.University, Wageningen

Subject Index

A
adsorption
 equilibrium 42
 non-equilibrium 42,43
approach
 Eulerian 4, 8
 Lagrangian 3
aquifer 16, 20, 22, 23, 24, 25, 26, 59, 61, 72, 81, 91, 140, 141, 156, 164, 165, 167,173, 182, 212, 213

B
boundary conditions 27, 32–33, 53, 55, 56, 57, 61,68, 83, 85, 90, 94, 98, 99–101, 111, 119, 120, 125, 127, 130, 135, 140, 156, 165, 183–184, 189, 210, 212, 213, 217, 218
 Cauchy 27
 Dirichlet 27
 main 56, 57, 94
 natural 56, 57, 94
 Neumann 27
bubbling pressure 30
bulk density 41, 122, 202, 206, 210

C
capillary
 conductivity 30
 diffusivity 32
 force 13, 47
 fringe 19, 28, 47, 48
 pressure 45
coefficient
 angular 12
 Fourier 52
 kinetic of the first order 44
 transport 41
 of capillary diffusivity 32
 of compressibility 16, 21
 of dispersion 36–39, 109, 111, 118
 of dispersivity 37, 206
 of distribution 42, 44, 120, 160, 202, 206, 210
 of dynamic viscosity 9, 10, 15, 46
 of effective porosity 157, 202, 206
 of hydraulic conductivity 15–17, 19, 21, 24, 25, 26, 31, 61, 93, 125, 132, 135, 137, 140, 141, 146, 151, 155, 156, 164, 182, 184, 201, 207, 210, 212, 213, 216
 of hydrodynamic dispersion 37
 of kinematic viscosity 10
 of molecular diffusion 36, 109, 209
 of permeability 15, 46
 of second viscosity 9
 of specific storativity 22, 24, 72
 of storativity 24, 65, 72
 of the rate of sorption 122
 of tortuosity 36
 of transmissivity 16, 24, 65, 69, 72, 93, 210
 of unsaturated hydraulic conductivity 30, 207, 209
critical capillary head 30
curve
 breakthrough 40, 44, 161
 characteristic 116
 envelope 11
 gradation 109
 pF 29
 retention 29, 30, 32

D
derivative
 local 5
 substantial 5, 6, 7
dispersivity
 longitudinal 37, 164, 202, 206
 tensor of 37
 transversal 37, 164, 202, 206
displacement 4, 9, 35

E
element
 boundary 88, 93, 114, 140, 155, 156, 192
 constant 89, 93
 hybrid 90
 interzone 92, 93, 140, 155

linear 89
  infinite 76, 77
  isoparametric 74, 77,78, 80, 106, 113, 126, 128, 129
  subparametric 74, 78, 81, 131
  superparametric 74
  tetrahedron 78
  triangular 72, 77, 78, 93, 119, 129, 146
equation
  Boussinesq 24, 25
  capacity form 31, 32
  constitutional 9
  Darcy-Buckingham 30
  diffusion form 31
  Dupuit 213
  Euler 7, 10
  Fokker-Planck 118
  Laplace 11, 20, 56, 57, 85, 94
  mass continuity 6
  Navier-Stokes 9–10
  of continuity 10, 20, 21, 39, 46
  of force's equilibrium 7
  of steady groundwater flow 20
  of streamline 11
  of the transport 40
  Poisson 25, 91, 100, 114
  Richards 102
equipotential lines 11, 12

F
field capacity 30
flow
  isothermal 10
  laminar 18
  mass 5, 36, 37, 170
  multiphase 45, 46
  potential 10–12, 19
  turbulent 18
fluid
  incompressible 10, 23
  Newtonian 9, 10
force
  capillary 13, 47
  inertial 7, 18
  molecular 13, 45
  surface 7
  volumetric 6
formula
  Dupuit 69
  Forchheimer 18
  Green 57
  Poiseuille 15, 18
  recurrent 71, 104, 112, 115

functions
  base 52–57, 70, 71, 72, 74, 75, 77, 78, 80, 84, 85, 87, 89, 90, 93, 98, 103, 104, 105, 106, 110, 113, 121, 126, 129
  Bessel, modified 96
  distribution 111
  mapping 74, 76
  weight 53–57, 70, 85, 87, 88, 92, 94, 110, 113, 121, 193, 197

H
head
  piezometric 14, 17, 20, 23
  pressure 17, 29, 31, 32, 103, 104, 106, 146, 207, 209, 210
  velocity 17
hysteresis 29, 31, 102

I
isotherm
  Freundlich 43, 207
  Langmuir 43
  linear 42, 43, 119, 120, 121

L
law
  associative 50
  commutative 50
  Darcy 16–19, 20
  Darcy-Buckingham 31
  Darcy-Lewerett 46
  Fick 36–39
  Hooke 21

M
method
  ADI 68
  boundary element 58, 59, 84–100, 114, 119, 175, 201, 202
  boundary integral 59
  characteristic curves 110, 116–117, 129
  Choleski 112
  collocation 54, 106
  DRM 100–101, 113–116, 126, 164
  finite differences 58, 64–69, 82, 102, 110, 117, 119, 126, 125, 127
  finite element 51, 55, 59, 69–84, 85, 103, 111–113, 116, 119, 120–123, 126, 131, 207
  Galerkin 54, 57, 106, 113
  Gauss elimination 189
  iteration 68, 104
  moments 53
  Newton-Raphson 104

Petrov-Galerkin 110, 112, 122, 130
random walk 110, 117–119, 121, 123, 126, 129, 130, 153, 175, 201
Ritz 55
skyline 83
weighted residuals 52–55, 85, 109
model
   3-D 61, 78, 127, 131, 145, 151, 164, 165
   mathematical 1, 3, 30, 125, 144
   numerical 1, 145
   one-site 44
   planar 77, 126, 153, 175
   two-dimensional 23
   two-site 44

N

number
   Peclet 109, 112, 117, 118
   Reynolds 18

P

porosity 20, 31, 46, 123
   aerated 15
   effective 15, 30, 40, 43, 122, 164, 206
   volumetric 14
potential
   complex 12, 62, 63
   Girinski 25–27, 62, 63
   gravity 13, 28
   groundwater 17, 28
   moisture 28
   osmotic 13
   pressure 13, 207, 209, 217
   scalar 10
   total 13
   velocity 11

R

retardation factor 43, 120, 121, 122, 160

S

solution
   analytic 61–64, 102, 111, 120, 186, 197
   approximate 70, 85, 91
   frontal 82, 128
   fundamental 58, 86, 92, 94, 96, 97
   iterative 211

particular 100
weak 55–57, 70, 85, 103, 111
space 49–50
   C 51
   L^2 51, 53, 56, 58
   linear 49
   metric 50
   normed 50
   Sobolev 51, 53, 55, 56, 57, 71, 76, 87, 105
   unitary 50
streamlines 11, 63, 140, 141, 142, 156, 196, 197, 200, 213, 214

T

theorem
   Dupuit 23
   Gauss-Ostrogradski 7
   Green 56, 57
   Reynolds transport 6
transport
   advective 40, 109, 116, 117, 130
   disperse 112, 116, 117, 118
   of pollutants 35, 36, 39, 40, 43, 45, 48, 123, 157, 164
   one-dimensional 37, 111

V

velocity
   complex 12
   porous 15, 20, 37, 40, 121, 157, 197

W

water
   capillary 13
   crystallic 13
   gravitational 13, 15
   pellicular 13, 19, 40
water capacity 31

Z

zone
   aeration 19
   saturated 13, 16, 19, 28, 30, 31, 32, 40, 43, 61, 70, 102, 122, 125, 207
   unsaturated 13, 28, 30, 32, 40, 45, 48, 101, 102, 103, 106, 164, 207, 209, 217

Druck: Strauss Offsetdruck, Mörlenbach
Verarbeitung: Schäffer, Grünstadt

Nutzungs- und Garantiebedingungen

§ 1 Vertragsabschluß
Durch Öffnen der CD-ROM-Hülle vereinbart der Endnutzer mit dem Springer-Verlag die Nutzungs- und Garantiebedingungen, die auf der gegenüberliegenden Seite abgedruckt sind. Falls der Endnutzer dies nicht anerkennen will, kann er die ungeöffnete Packung mit dem Original-Kaufbeleg binnen zwei Wochen gegen volle Erstattung des Kaufpreises seinem Lieferanten oder dem Springer-Verlag zurückgeben. Für die Rückgabe gilt § 7.

§ 2 Urheber- und Nutzungsrechte
1. Alle Nutzungsrechte an der Software (Programme und Quellcode) stehen dem Autor zu. Die Software ist urheberrechtlich geschützt.
2. Der Springer-Verlag überläßt dem Endnutzer die nicht ausschließliche schuldrechtliche Befugnis, die Software vertragsgemäß zu nutzen. Vertragsgemäß ist nur eine Nutzung, bei der das Programm mit Hilfe der beschriebenen Anweisungen ausgeführt wird. Insbesondere sind das Verändern, Bearbeiten, Umgestalten und Decompilieren der Software unzulässig.
3. Das Programm darf zur selben Zeit nur auf einem Rechner und auf einem Arbeitsplatz benutzt werden. Bei Nutzung auf Rechnern mit mehreren Arbeitsplätzen oder in Netzen muß pro Arbeitsplatz, auf dem die Nutzung möglich ist, eine Lizenz erworben werden.
4. Das Programm darf nur einmal zu Sicherungszwecken vervielfältigt werden.

§ 3 Weitergabe der Software
1. Jede Weitergabe (z.B. Verkauf) der Software an Dritte und damit jede Übertragung der Nutzungsbefugnis und -möglichkeit bedarf der schriftlichen Erlaubnis des Springer-Verlages oder des Autors.
2. Der Springer-Verlag wird die Erlaubnis geben, wenn der bisherige Endnutzer dies schriftlich beantragt und eine Erklärung des nachfolgenden Endnutzers vorliegt, daß dieser sich an die Regelungen dieses Vertrages gebunden hält. Ab dem Zugang der Erlaubnis erlischt das Nutzungsrecht des bisherigen Nutzers und wird die Weitergabe zulässig.

§ 4 Unerlaubte Nutzung
1. Die gesamte Software ist durch Urheberrecht, Warenzeichenrecht, Wettbewerbsrecht und diesen Vertrag geschützt. Verstöße hiergegen können zivilrechtlich und strafrechtlich verfolgt werden.
2. Der Käufer haftet dem Springer-Verlag für alle Schäden und Nachteile aufgrund von Verletzungen dieser Regelung.

§ 5 Funktionsbeschränkungen der Software
1. Nach dem Stand der Technik können Fehler der Software auch bei sorgfältiger Erstellung nicht ausgeschlossen werden.
2. Die Software dient zum Auswerten und Visualisieren von Daten.
3. Für die Funktionsfähigkeit des Programmes sind die im Buch beschriebene Hardware und Basissoftware notwendig. Die Installation der Software muß genau nach den Vorschriften erfolgen. Abweichungen hiervon können zu Schäden auch an der Hardware, an anderer Software und an Daten führen.

§ 6 Garantie
1. Bei berechtigten Beanstandungen hat der Springer-Verlag zunächst die Möglichkeit, dem Endnutzer ein anderes Exemplar zu überlassen (auch ein anderes Programm-Release). Wenn damit die Beanstandung nicht behoben ist, kann der Endnutzer von seinem Lieferanten den Kaufpreis zurückverlangen, wenn er die Software entsprechend § 7 zurückgibt.
2. Die Inanspruchnahme der Garantie setzt voraus, daß der Endnutzer den Mangel schriftlich genau beschreibt.
3. Auf Minderung und Nachbesserung hat der Endnutzer keinen Anspruch. Im übrigen gelten die Regeln der kaufrechtlichen Gewährleistung (§§ 459 - 480 BGB) entsprechend.

§ 7 Rückgabe
1. Der Kunde kann die Software (z.B. nach § 1 oder § 6 Abs. 1) nur komplett und mit dem Original-Kaufbeleg zurückgeben. Zusätzlich hat er die Erklärung abzugeben, daß keine Kopien existieren.

§ 8 Beratung
1. Der Springer-Verlag eröffnet die Möglichkeit, Fragen in bezug auf die Software an den Autor zu richten. Ein Rechtsanspruch für diesen Dienst besteht jedoch nicht.
2. Die Fragen können die Installation, die Handhabungs- und Benutzungsprobleme des Programms betreffen.
3. Anfragen sind schriftlich oder über Mailbox an den Springer-Verlag zu richten. Der Springer-Verlag vermittelt lediglich ungeprüft die Beantwortung durch den Autor. Die Antworten erfolgen üblicherweise in der Reihenfolge des Eingangs. Nicht jede Frage wird beantwortet werden können.

§ 9 Haftung
1. Der Springer-Verlag und der Autor haften nur bei Vorsatz, bei grober Fahrlässigkeit und Eigenschaftszusicherungen. Die Zusicherung von Eigenschaften bedarf der ausdrücklichen schriftlichen Erklärung. Für Auskünfte nach § 8 wird nicht gehaftet.
2. Die Haftung aus dem Produkthaftungsgesetz bleibt unberührt.
3. Der Einwand des Mitverschuldens des Endnutzers bleibt dem Springer-Verlag offen.

§ 10 Schluß
1. Gerichtsstand für alle Klagen im Zusammenhang mit der Software und dieser Vereinbarung ist D-69115 Heidelberg, wenn der Vertragspartner Vollkaufmann oder gleichgestellt ist oder keinen allgemeinen Gerichtsstand in Deutschland hat.
2. Es gilt ausschließlich das Recht der Bundesrepublik Deutschland mit Ausnahme der UNCITRAL-Kaufgesetze.
3. Sollte eine Bestimmung dieses Vertrages unwirksam sein oder werden oder sollte der Vertrag unvollständig sein, so wird der Vertrag im übrigen inhaltlich nicht berührt. Die unwirksame Bestimmung gilt als durch eine solche Bestimmung ersetzt, welche dem Sinn und Zweck der unwirksamen Bestimmung in rechtswirksamer Weise wirtschaftlich am nächsten kommt. Gleiches gilt für etwaige Vertragslücken.

Conditions of use and terms of warranty

This is an unauthorized translation of the original text in German language. Only the original German text is legally binding.

§ 1 Concluding the contract
Opening the sealed plastic cover binds the end user to the conditions of use and the terms of the warranty. If the end user does not wish to be bound by these conditions, he should return the unopened package to his supplier or to Springer-Verlag and the selling price will be refunded. For the return of goods, § 7. is valid.

§ 2 Copyright and conditions of use
1. All rights pertaining to the Software (program and source code) are owned by the author. The Software is protected by copyright.
2. Springer-Verlag grants the end user, subject to legal liability, the non-exclusive right to use the Software as described by the terms of this contract. Under this contract use of the program is restricted to that carried out according to the instructions described. The decompiling, disassembling, reverse engineering or in any way changing the program is expressly forbidden.
3. The program may, at any one time, only be used on one computer at a single workplace. When used on computers with several or many terminals or in a network, a license application must be made for each workstation or terminal on which use is possible.
4. The program may be copied once for backup purposes.

§ 3 Transfer of the Software
1. Any transfer (e.g. sale) of the Software to a third party and with it the transfer of the right and the possibility of its use may only occur with the written permission of Springer-Verlag or the author.
2. Springer-Verlag will give this permission when the end user up to this point makes a written application and the subsequent end user makes a declaration that he will remain bound by the terms of this contract. Receipt of permission terminates the right of the first end user to operate the program and the transfer to the second end user may take place.

§ 4 Unauthorized use
1. The complete Software is protected by the laws of copyright, the laws governing the use of trademarks, the laws of trade and commerce this contract. Violations may lead to action being taken under civil and criminal law.
2. The buyer is liable to Springer-Verlag for any damages or detriment accruing from any infringement of these regulations.

§ 5 Functional limitations of the Software
1. Even with the lates state of technological development and with meticulous care being taken during production, errors in the Software cannot be excluded.
2. The Software evaluates and visualizes data.
3. The hardware and basis software described in the book are necessary for the functional capability of the program. The installation of the Software must be carried out exactly as described in the instructions. Deviation from these instructions can lead to damage of the hardware and also to other software and data.

§ 6 Warranty
1. In response to justified claims, Springer-Verlag has, as first possibility, the option of supplying the user with another copy of the program (including another program release). If the claim is still not remedied, the end user can demand the return of the selling price from his supplier when he returns the Software in compliance with the terms set out in § 7.
2. A prerequisite to making a claim under the warranty is that the end user supplies an exact description of the defect in writing.
3. The end user has no claim to a reduction in the selling price or to correction of defects. In other respects the German Code of Civil Law (BGB) concerning the warranty of goods shall apply (§§ 459 to 480 BGB).

§ 7 Returning the software
1. The customer can only return the Software (e.g. according to § 1 or § 6 Sect. 1.) in its entirety together with the original sales receipt/invoice. In addition he has to hand over the declaration stating that no copies remain in his possession.

§ 8 Help
1. Springer-Verlag has inaugurated the possibility of asking the author questions with reference to the Software. However, this is a voluntary service and is not the customer's right.
2. The questions can be concerned with installation, operation, and problems of utilization.
3. Questions should be mailed or sent via mailbox to Springer-Verlag (see handbook for instructions). The answers from the author are merely forwarded by Springer-Verlag without being checked. The questions are normally answered in the order they are received. It will not be possible to answer every question.

§ 9 Liability
1. Springer-Verlag and the author are only liable for willful intent, gross negligence, and when the program fails to fulfill its assured purpose and function. The assured purpose and functions are those which are explicitly declared in writing. There is no liability for information described in § 8.
2. The liability under German law for product liability is unaffected.
3. The plea that the end user is also at fault remains an option for Springer-Verlag.

§ 10 Conclusion
1. The location of the competent court for all legal action in connection with the Software and this contract is D-69115 Heidelberg if the contract partner is a registered trader or equivalent, or if he has no legal domicile in Germany.
2. This contract is exclusively governed by the laws of the Federal republic of Germany with the exception of the UNCITRAL laws of trade and commerce.
3. Should any provision of the contract prove unenforceable or if the contract is incomplete, the remaining provisions will remain unaffected. The invalid provision shall be deemed replaced by the provision which in a legally binding manner comes nearest in its meaning and purpose to the unenforceable provision. This shall apply to any omission in the contract that may occur.